What Makes Variables Random

Probability for the Applied Researcher

What Makes Variables Random

Probability for the Applied Researcher

Peter J. Veazie, PhD

CRC Press
Taylor & Francis Group
Boca Raton London New York

CRC Press is an imprint of the
Taylor & Francis Group, an **informa** business

A CHAPMAN & HALL BOOK

CRC Press
Taylor & Francis Group
6000 Broken Sound Parkway NW, Suite 300
Boca Raton, FL 33487-2742

First issued in paperback 2020

© 2017 by Taylor & Francis Group, LLC
CRC Press is an imprint of Taylor & Francis Group, an Informa business

No claim to original U.S. Government works

ISBN 13: 978-0-367-57371-3 (pbk)
ISBN 13: 978-1-4987-8108-4 (hbk)

Library of Congress Cataloging-in-Publication Data

Names: Veazie, Peter J.
Title: What makes variables random : probability for the applied researcher / Peter J. Veazie.
Description: Boca Raton : CRC Press, 2017. | Includes bibliographical references.
Identifiers: LCCN 2016057398 | ISBN 9781498781084 (hardback)
Subjects: LCSH: Random variables. | Variables (Mathematics) | Probabilities.
Classification: LCC QA273 .V38 2017 | DDC 519.2--dc23
LC record available at https://lccn.loc.gov/2016057398

Visit the Taylor & Francis Web site at
http://www.taylorandfrancis.com

and the CRC Press Web site at
http://www.crcpress.com

To Wendy, Matthew, and Devin

Contents

Section III   Applications

Preface

A number of years ago, I noticed that the growing popularity of methods such as hierarchical modeling was accompanied by a pattern of misuse. For example, researchers were using these methods to assure appropriate standard errors, but in doing so some were confusing the statistically meaningless concept of nested data with the statistically relevant concept of a nested data generating process. This is a misunderstanding that can lead to the misapplication of the methods. As these methods became more common, so did their misuse. The underlying problematic issue arises more generally in statistical analysis when data are confused for the data generating process, and variables defined on the data are confused for random variables defined on the data generating process.

Considering the source of this confusion, I settled on what would become the title of this book: there seemed to be a lack of understanding regarding what makes variables random. Distinguishing data from the process that produced it is essential to understanding statistics as a tool for empirical analysis.

Having identified this problem, what was the solution? As many applied researchers do not have a mathematical background beyond calculus, I tried to formulate the necessary understanding of random variable in terms of a calculus-based framework. Unfortunately, this approach seemed inadequate: I was unable to use calculus alone to provide the conceptual depth required to get at what really makes variables random.

I turned to measure theory. However, I was aware that many would not be familiar with measure theory; indeed many would not have a background in real analysis. Moreover, many would likely, and rightfully, not be interested in developing such a mathematical background. I wondered whether measure theory and probability could be taught at a level sufficient to provide a conceptual tool without having to resort to the depth required for a mathematical tool. Would a measure-theoretic conceptualization in conjunction with college-level calculus be sufficient for applied researchers to better understand and better use the statistical methods with which they were already familiar? A number of years ago, I presented an 8-hour workshop focused on providing an affirmative answer to this question. Although the workshop was a successful introduction, the timeframe was insufficient to provide either the depth or scope necessary for the impact I was seeking.

Following the workshop, I expanded its notes into the book presented here. My goal was to produce a short text to augment a researcher's existing calculus-based understanding of probability. My hope is that the resulting book achieves its purpose by providing a measure-theoretic conceptualization of probability and its use in informing research design and statistical analysis.

I greatly appreciate the feedback from those who read draft sections of the text or listened to me as I incorporated its concepts into lectures on statistical methods. I am particularly indebted to Viji Kannan at the University of Rochester for her willingness to challenge the content and clarity of this text: Our discussions of measure theory and its application helped shape my thoughts about this project, its content, and its presentation.

Section I

Preliminaries

1

Introduction

For the applied researcher, mathematical probability is a tool—a means to investigate real-world phenomena. As such, many researchers learn and understand this tool in a language that facilitates direct utilization, often in terms of calculus as taught to undergraduates. Unfortunately, a strictly undergraduate-level calculus-based understanding does not always provide a sufficiently rich conceptual framework by which mathematical probability can be connected to real phenomena. Consequently, the applied researcher may engage in analysis of data without knowing what mathematical probability and statistics are representing in their investigation, thereby risking a mistaken interpretation. For example, analysts often speak of "nested data" to refer to the structure in the data they presume informs the analysis; however, nested data is not a statistically meaningful concept but rather a misconception that can be avoided with a proper understanding of probability.

Mathematicians, statisticians, and a few other disciplines will have learned mathematical probability in terms of measure theory. Not only does this mathematical perspective give them great power in constructing careful proofs in probability theory, it also provides the conceptualization that facilitates an easy translation between real-world problems and mathematical probability. Unfortunately, using measure theory operationally can be overkill for empirical analysis: understanding measure theory with sufficient mathematical rigor to use it for "doing the math" is rightfully deemed a waste of time for many applied researchers. And so, the conceptual baby is thrown out with the operational bathwater. Authors on the subject tend not to integrate the two for applied researchers. Either they conceptualize and operationalize with measure theory or they conceptualize and operationalize with standard calculus. Aris Spanos' text *Probability Theory and Statistical Inference* is a rare and excellent exception.

However, in view of the numerous calculus-based texts on probability and statistics that already exist, as well as the training many researchers may already have, another such comprehensive integrative text is not needed. What is needed is a brief text that provides a basic conceptual introduction to measure theory, probability, and their implications for applied research. Consequently, it is the goal of this short text to augment the applied researcher's existing calculus-based understanding of probability: to generate a measure-theoretic conceptualization of mathematical probability such that researchers can better use their calculus-based probability framework to design studies, analyze data, and appropriately understand results.

TABLE 1.1

Hospital Data

Physician	Patient	HbA1c	Pt Age
Harriet	James	7.7	67
Harriet	Mary	6.8	62
Harriet	John	7	73
Fred	Robert	8.3	88
Fred	Patricia	9.1	66
Lisa	Linda	9.2	86
Lisa	Barbara	7.5	63
Lisa	Michael	8.3	71
Lisa	Elizabeth	6.7	77

To achieve this goal and to make this text useful for those without the in-depth mathematical background who seek a broad-level understanding, I necessarily compromise on mathematical depth and detail. For those interested in a more mathematically rigorous treatment of the topic, see the references in the "Additional Readings" section at the end of Chapter 4, among others that can be found by searching on the terms *measure theory* and *probability*.

Imagine that you are approached by the CEO of a hospital, who wants you to evaluate the performance of the doctors in the hospital. Suppose she hands you the data in Table 1.1. The data contain patient and physician identifiers and measures for patient age and glycosylated hemoglobin (HbA1c) levels (lower numbers are better in the HbA1c range reported here).

The average HbA1c across these patients for Dr. Harriet is approximately 7.2, the average across these patients for Dr. Fred is approximately 8.7, and the average for Dr. Lisa is approximately 7.9. Dr. Harriet's patients in the data have better control of their HbA1c on average than the other physicians' patients in the data, and Dr. Fred's patients are worse on average than the other physicians' patients. You already know how to calculate these basic statistics. This book will help you understand and answer more nuanced questions such as the following:

1. Would clustering by physician be appropriate?
2. What do reported standard errors of statistics mean?
3. Would a random effects, or multilevel, model be appropriate?

The answers to these questions are the same: "there is not enough information to tell." This is because the data alone do not provide sufficient information to identify or understand the meaning of statistics as used in applied research. By the end of this book, you will understand why and be able to better design your analyses and understand your results.

Thus far, I have been careful to use the phrase *mathematical probability* rather than the term *probability*. The distinction is extremely important throughout the conceptual development in this book. The term *probability* is ambiguous. It is sometimes used to represent a mathematical structure, which only implies strict mathematical results, and it is sometimes used to mean a specific real-world concept such as a source of uncertainty. Although such substantive interpretations of probability are necessary to the work of applied research, it should not be confused with the implications of the mathematics. In this text, I encourage thinking of mathematical probability as a model of a substantive phenomenon. However, once a model is properly developed to capture a phenomenon of interest, its properties follow solely from the mathematics. The importance of this distinction will become clear throughout the book. Nonetheless, to avoid cumbersome overuse of the phrase *mathematical probability*, I will use the term *probability* in the rest of this book when the context makes it clear to which sense I am appealing.

The first section of this book comprises this introduction and a chapter reviewing set theory and functions, which may be skipped by readers who are familiar with those concepts. The second section of the book focuses on the basics of measure theory and probability. The third section focuses on the implications of measure theory to applied research—the use of a calculus-level mathematical understanding of probability informed by a measure-theoretic conceptualization.

Although I use the language of mathematics to communicate the requisite ideas, the goal of this text is to achieve a conceptual understanding of basic principles that will allow you to clearly think through research problems. I have written this book for researchers and students who already have an understanding of applied statistics in terms of undergraduate calculus. I therefore assume the reader has an understanding of probability, distributions, and statistics.

Additional Readings

The book by Aris Spanos titled *Probability Theory and Statistical Inference: Econometric Modeling with Observational Data* (Cambridge University Press, 1999) provides an introduction to probability and statistics that uses measure theory as a conceptual framework but uses basic calculus for its implementation. The book introduces probability, statistics, and statistical inference at a level useful for applied research. The intended audience of Spanos' book comprises those who have had at least a one-semester course in calculus.

For references to books on substantive theories of probability, see the "Additional Readings" section at the end of Chapter 5.

2

Mathematical Preliminaries

Before describing measure theory and mathematical probability, it would be helpful to review some basics of set theory and functions. This chapter covers the definitions and notation required to understand the remainder of the book. Readers who are familiar with set theory and functions can skim or skip this chapter without loss of continuity.

Set Theory

The essentials of set theory required to understand measure theory's relevance to applied research are captured by the following definitions.

A *set* is a collection of distinct objects (concrete or abstract). To compose a set, each object in the collection is distinct, and any object is either definitely in or definitely out of the set. The distinctiveness of the objects means that a collection of words such as {Mary, Fred, Mary} is a set that can also be represented, more efficiently, as the set {Mary, Fred}. This is because the former representation contained identical copies of the word *Mary*, whereas the latter does not, yet each contains the same distinct words. Note that it is common to use "curly" brackets (i.e., braces) to enclose members of a set.

The elements of a set need not be real; they may be imaginary, conceptual, or simply asserted to exist, as in "let A be a set of objects" or even more concisely as in "let A be a set," without further specification. It is important to remember that sets may have sets as elements, or sets of sets as elements, or sets of sets of sets as elements—you get the idea. For example, consider that the following are different sets: {1, 2, 3, 4, 5}, {{1}, {2}, {3}, {4}, {5}}, and {{{1}}, {2}, {3}, {4}, {5}}. The first set is the set of integers from 1 to 5; the second is the set of sets containing integers from 1 to 5. The third differs from the second in that the first listed member is the set containing the set that contains 1.

If an object a is one of those that compose a set A, we say "a is a *member* of set A" or "a is an *element* of set A." We denote the relation of "is a member of" or "is an element of" by the symbol $\in$. The claim that an object is not a member of a set is denoted by the symbol $\notin$. Consequently, the statement "a is a member of set A" is written as $a \in A$, and the statement "a is not a member of set A" is written as $a \notin A$.

If each member of a set B is also a member of a set A, then we can say that set B is *contained* in set A or that set A *contains* set B. In this case, B is considered a *subset* of A. Moreover, if B is contained in A, and A has at least one element that is not a member of B, then B is a *proper subset* of A. If there is no such remaining element in A (i.e., B is contained in A and A is contained in B), then A and B are considered *equal sets*. The symbol $\subseteq$ is used to denote "subset of;" for example, $B \subseteq A$ means that B is a subset of A. The symbol $\subset$ is used to denote "proper subset of;" for example, $B \subset A$ means that B is a proper subset of A, which implies that all members of B are also members of A, but there is at least one member of A that is not a member of B. The symbol $\approx$ is used to denote equal sets, as in $A \approx B$, indicating that sets A and B are equal. If it is the case that a set A is contained in another set B, and that set B is also contained in set A, then sets A and B are equal (i.e., if $A \subseteq B$ and $B \subseteq A$, then $A \approx B$). This should be evident because if A is contained in B, then B contains all elements that are in set A, but if B is also contained in set A, then A contains all elements that are in set B. Consequently, there does not exist an element in one of the sets that is not also an element of the other.

Note the distinction between the concepts "member of" (or "element of") and "subset of." The former identifies a particular object that is a component of a set, whereas the latter identifies a set whose elements are also elements of another set. For example, regarding set A defined as $\{1, 2, 3, 4, 5\}$, we can properly say that the number 2 is an element of A, and the set with 2 as its single element is a subset of A (i.e., $2 \in A$ and $\{2\} \subset A$). However, for set B defined as $\{\{1\}, \{2\}, \{3\}, \{4\}, \{5\}\}$, we can properly say that the set with 2 as its single element is an element of B, the set containing the set that contains the number 2 is a subset of B, and the number 2 itself is neither an element of B nor a subset of B (i.e., $\{2\} \in B$ and $\{\{2\}\} \subset B$, but $2 \notin B$ and $2 \not\subset B$). With respect to A, the number 2 is a member and the set $\{2\}$ is a subset; however, for set B the set $\{2\}$ is a member, and therefore the set that contains the set $\{2\}$ (i.e., $\{\{2\}\}$) is a subset of B.

In set theory, there exists a particular set that contains no members at all. This set is called the *empty set*; it is commonly denoted as $\varnothing$. It is typically a matter of mathematical convention to consider the empty set to be a subset of every set; this convention is adhered to in this book.

A set with only one element is called a *unit set* or *singleton*. For example, $\{2\}$ is a singleton containing the number 2, and $\{\{Fred, Lisa\}\}$ is a singleton containing the set $\{Fred, Lisa\}$.

The set of all subsets of a given set A is called the *power set* of A, denoted as $\wp(A)$. For example, the power set of $A = \{1, 2, 3\}$ is $\wp(A) = \{\{1\}, \{2\}, \{3\}, \{1, 2\}, \{1, 3\}, \{2, 3\}, \{1, 2, 3\}, \varnothing\}$. Note that the empty set is included because, as stated above, it is considered a subset of all sets, which means it is a subset of A and therefore belongs in the power set of A.

We use curly brackets to explicitly list the members of sets. We also use them to represent a set by denoting an arbitrary member and a rule by which the members are defined; the arbitrary member and rule are separated by a colon, which can be read as the phrase "such that." For example, $\{w: w \in A\}$ denotes the set of elements w such that w is a member of set A; or another example, $\{w: w \in A$ and $w \notin B\}$ denotes the set of elements w such that w is a member of set A and not a member of set B. With this notation in hand, we can define operations on sets (new notation for these operations is introduced in the definitions).

Let A and B be sets; then their *union* (denoted by the symbol $\cup$) is the set of elements that belong to either A or B or both:

$$A \cup B = \{w: w \in A \text{ or } w \in B\} \tag{2.1}$$

For example, if set A is {Fred, Lisa, Bill, Sue}, and set B is {Bill, Sue, Henry, Linda}, then the union of A and B is {Fred, Lisa, Bill, Sue, Henry, Linda}: the set that includes all members of each of A and B. Notice that because sets do not contain redundant labeling, even though Bill and Sue are elements of both A and B, they show up only once in the set that composes the union.

Let A and B be sets; then their *intersection* (denoted by the symbol $\cap$) is the set of elements that belong to both A and B:

$$A \cap B = \{w: w \in A \text{ and } w \in B\} \tag{2.2}$$

For example, if set A is {Fred, Lisa, Bill, Sue} and set B is {Bill, Sue, Henry, Linda}, then the intersection of A and B is {Bill, Sue}, the set that includes all the members that sets A and B share. For a graphical representation of a union and intersection of sets, see Figure 2.1.

A different notation is often used to indicate unions and intersections across indexed collections of sets. For example, consider the collection of K sets $\{A_1, A_2, \ldots, A_K\}$. Rather than listing each set connected with a union or intersection symbol, such as $A_1 \cup A_2 \cup \ldots A_K$ and $A_1 \cap A_2 \cap \ldots A_K$, a simpler notion is often adopted:

$$A_1 \cup A_2 \cup \ldots A_K = \bigcup_{k=1}^{K} A_k \tag{2.3}$$

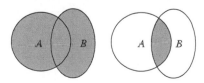

FIGURE 2.1

Union and intersection: the shaded area of the left-hand graphic depicts the union of sets A and B, whereas the shaded area of the right-hand graphic depicts the intersection of sets A and B.

and

$$A_1 \cap A_2 \cap \ldots A_K = \bigcap_{k=1}^{K} A_k \tag{2.4}$$

For a set B, the *complement* of B (denoted by a bar placed above B) is the set of elements that are not a member of B:

$$\overline{B} = \{w : w \notin B\} \tag{2.5}$$

Figure 2.2 presents a graphical representation of the complement of a set.

The complement of a set presupposes a *basic* or *universal* set U such that $U = B \cup \overline{B}$. Without the context of the universal set, it is difficult to identify what exactly is the complement of a set. For example, if B is defined as the set of siblings of a given family {Fred, Lisa, Bill, Sue, Henry, Linda}, what is $\overline{B}$? Is it the rest of the immediate family, all the rest of their living relatives, all people alive, all people who ever lived, all other physical objects in the world, or all physical objects plus the concepts of *liberty, peace,* and *blue*? Clearly, a set is understood by the content of its members; the complement of a set, which is itself a set, must similarly be understood and thereby necessitates a basic or universal set. This is clearer if the universal set, say U, is included in the definition, such as $\overline{B} = \{w : w \in U \text{ and } w \notin B\}$. For example, Figure 2.2 depicts the complement of set B as relative to the rectangle containing it, as opposed to, say, the page on which it is drawn.

Two sets A and B are considered to be *disjoint* if their intersection is the empty set, which is to say that sets A and B do not share any members:

$$A \cap B = \emptyset \tag{2.6}$$

For example, for A defined as {Fred, Lisa, Bill, Sue} and B defined as {Henry, Linda}, the intersection of A and B is empty because neither set contains a member of the other: they are disjoint. Figure 2.3 presents a graphical representation of disjoint sets.

A collection of k sets $A_1, A_2, \ldots, A_k$ is considered to be a *disjoint collection* of sets if each distinct pair of sets in the collection are disjoint by the preceding definition. For example, consider four sets: A_1 defined as {Fred, Lisa}, A_2 defined as {Bill, Sue}, A_3 defined as {Henry}, and A_4 defined as {Linda}.

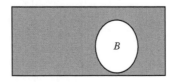

FIGURE 2.2
Complement: the shaded area represents the complement of the set B.

FIGURE 2.3
Disjoint sets: sets A and B do not overlap and therefore do not share any elements: they are disjoint.

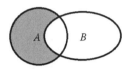

FIGURE 2.4
Relative complement: the shaded area in this graphic depicts the relative complement of the set B with respect to the set A.

None of these sets share elements with any other; they are a disjoint collection of sets.

The set of elements that are members of set A but are not members of set B is the *relative complement* of B with respect to A, sometimes called the *set difference* of the set A with respect to B (denoted by either a backslash \ or a minus −):

$$A \backslash B = A - B = A \cap \overline{B} = \{x : x \in A \text{ and } x \notin B\} \tag{2.7}$$

For example, if set A is {Fred, Lisa, Bill, Sue} and set B is {Bill, Sue, Henry, Linda}, then the relative complement of B with respect to A is {Fred, Lisa} and the relative complement of A with respect to B is {Henry, Linda}. Figure 2.4 presents a graphical representation of the relative complement.

A disjoint collection of nonempty sets $A_1, A_2, \ldots, A_k$ such that $S = \cup_{i=1}^{k} A_i$ is called a *partition* of the set S. In other words, if you chop up a set S into subsets that are mutually exclusive (i.e., each member of the set S can only be in one of the subsets) and exhaustive (i.e., each member of the set S must be in one of the subsets), then this collection of subsets is a partition of the set S. For example, if set A is {Fred, Lisa, Bill, Sue}, then the three sets $A_1 = $ {Fred, Bill}, $A_2 = $ {Lisa}, and $A_3 = $ {Sue} are a disjoint collection of sets that constitute a partition of A. Figure 2.5 presents a graphical representation of a partition.

For two partitions π_0 and π_1 of a set S, π_1 is a *refinement* of π_0 if each set in π_1 is a subset of one in π_0 and at least one set in π_1 is a proper subset of one in π_0. The partition π_1 is considered to be finer than π_0, and π_0 is considered to be coarser than π_1. Essentially, a refinement of a partition π_0 is achieved by partitioning at least one of its member sets into subsets.

A sequence of sets (e.g., $A_1, A_2, \ldots, A_k$) in which each is a proper subset of the preceding one (e.g., $A_1 \supset A_2 \supset \ldots \supset A_k$) is a *nested* sequence of sets, as

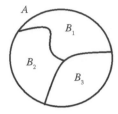

FIGURE 2.5
Partition: the sets B_1, B_2, and B_3 compose a partition of the set A (the circle).

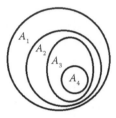

FIGURE 2.6
Nested sets: sets A_1, A_2, A_3, and A_4 compose a collection of nested sets centered on A_4 (or any subset of A_4).

shown in Figure 2.6. A sequence of such sets that can be indexed to the whole numbers, which continue infinitely, is an infinite nested sequence of sets. If the intersection of a collection of sets contains a set A, then the collection of sets is considered to be *centered* on set A. Consequently, a nested sequence of sets is centered on any subset of the last set in the sequence.

I will ostensibly define *continuous set* to mean a set such as the real line, intervals on the real line, areas of a plane, three-dimensional volumes, and higher-dimensional hyper-volumes. A characteristic of such a set is that no matter how close an inspection that you give around a point in the set, there is an infinite number of points in that region under inspection. This is a cumbersome, and pedestrian, characterization of a continuous set. Indeed, this use of the phrase *continuous set* is not true to the usual parlance of mathematics. The word *continuous* is better reserved for describing functions; however, to properly describe the notion I am presenting would require an understanding of metric spaces, limit points, and isolated points, or an understanding of sets that can support a properly defined continuous function. Given the limited reference I will make to these sets (whatever we wish to call them), it is not worth the conceptual effort to be proper. So, to restate, I will call a set that comprises a continuum of points a *continuous set*. In light of the preceding definitions, we can meaningfully speak of a nested sequence of continuous sets centered on some specified set. Figure 2.6 shows a set of nested sets centered on a set A_4.

Functions

A complete description of the mathematical concept of a function and its consequences is well beyond the scope of this book. For our purposes, we only need to have an understanding of some basics. However, before presenting a formal definition, let's consider some examples.

Example 2.1

Suppose I have two lists from an elementary school in which each student has exactly one teacher: one list contains the names of all students, the other list contains the names of all teachers. Suppose further that I identify for every student on the first list that student's teacher from the second list (see Figure 2.7). This student–teacher relationship is a function from the student list to the teacher list.

Example 2.1 is carefully constructed to highlight the main features of a function. To see what they are, consider the following contrasting examples.

Example 2.2

Suppose I have two lists from a middle school in which music is an elective that only some students take, and if a student takes music they have only one music teacher. The first list contains all the students of the school; the second list contains all the teachers. Now suppose I identify the student–music teacher relationship from the first list to the second list (see Figure 2.8). This relationship is not a function.

Unlike Example 2.1, which is a function, Example 2.2 contains students in the first list who do not have a music teacher identified in the second list.

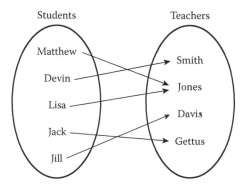

FIGURE 2.7
Function: the arrows represent a function that assigns each student to a teacher.

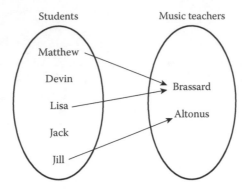

FIGURE 2.8
The arrows in this figure do not represent a function from students to teachers, because there exist some students who are not assigned a teacher. Specifically, Devin and Jack do not have a music teacher.

In order for a relationship to be a function, every member of the first list must be assigned to a member of the second list.

Example 2.3

Suppose I have two lists from elementary school: one contains the names of all students and the other contains the names of all first grade teachers. The student–teacher relationship is not a function in this case.

Example 2.3 appears subtly different from Example 2.2 in that the second list contains only a subset of teachers in the school. However, the result is essentially the same: The student–teacher relationship cannot identify a teacher for some of the students. And again, in order for a relationship to be a function, every member of the first list must be assigned to a member of the second list.

Example 2.4

Suppose I have a student list and a teacher list from a middle school in which some students have multiple teachers (see Figure 2.9). Identifying the teachers associated with each student does not constitute a function.

Example 2.4 highlights the fact that a function must assign only one element in the second list to each element in the first list (i.e., in this case a function must assign only one teacher to each student).

Example 2.5

Suppose I have a student list and an employee list from a given school. Identifying the principal associated with each student constitutes a function, even though each student is related to one and the same member of the employee list (see Figure 2.10).

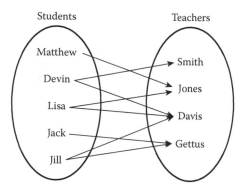

FIGURE 2.9

The arrows in this figure do not represent a function because some students (Devin, Lisa, and Jill) are assigned to more than one teacher.

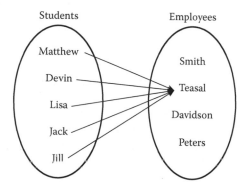

FIGURE 2.10

This figure represents a function because each student is identified with an employee—note that it does not matter that it is the same employee.

Whereas Example 2.4 did not exemplify a function because a function cannot identify each element of the first set (i.e., each student) to multiple elements of the second set (i.e., multiple teachers), Example 2.5 does exemplify a function because a function can identify multiple members of the first list (students) with the same member of the second list (employees).

A definition of a function sufficient to understand this text is as follows: Let X and Y be two sets, and let f be a relationship between the two sets that identifies one and only one member of Y with each member of X. Then f is a *function* from X to Y, written in this text as $f: X \rightarrow Y$. The set X is called the *domain* or *preimage* of function f, and the set Y is called the *codomain*. The member y of the range Y associated by function f with an arbitrary member x of the domain X is identified as $y = f(x)$. The set comprised of $f(x)$ for all x in X is called the *range* or *image* of the function f. Note the difference between the codomain and the range or image; this difference is shown in Figure 2.10. The set of employees is the codomain in this figure, whereas the set {Teasal} is the range or image.

It is important to note that the domain and codomain or range of a function are both sets, and that a set can itself contain sets. So Example 2.4 could be modified as in Example 2.6 below to achieve an appropriate function.

Example 2.6

Suppose we specify a set of students S = {Fred, Mary, Lisa, Greg} and a set of sets of teachers T = {{Mr. Smith}, {Ms. Johnson}, {Ms. Andersen}, {Mr. Smith, Ms. Johnson}, {Mr. Smith, Ms. Andersen}, {Ms. Johnson, Ms. Andersen}, {Mr. Smith, Ms. Johnson, Ms. Andersen}}. Now we can specify a function that identifies the set of teachers associated with each student. Perhaps Fred has both Smith and Johnson as teachers; our function works because the range T includes an element that is the set with both Smith and Johnson. Perhaps Mary has Mr. Smith, Ms. Johnson, and Ms. Andersen as teachers; again our function can work because the set comprising the three teachers is a member of T.

If the range of a function is equal to a set Z, then the function is said to map its domain *onto* Z (Figure 2.7 represents such a function), in which case the codomain and range are the same. If the range of a function is a proper subset of Z, then the function is said to map its domain *into* Z (Figure 2.10 represents such a function), in which case the codomain is larger than the range. For a function $f: X \rightarrow Y$, if $f(x) = f(w)$ implies $x = w$ and thereby each point in the range is associated with only one point in the domain, then f is a *one-to-one* function. If f is a one-to-one function, then there exists a function, say g, such that $g: Y \rightarrow X$ and $x = g(f(x))$ for all x in X. This function g is called an *inverse function* and is often labeled as f^{-1}, so that we would write $y = f(x)$ and $x = f^{-1}(y)$. Figure 2.11 presents an example of a one-to-one function.

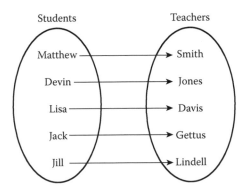

FIGURE 2.11
This figure represents a one-to-one function. This is a function from students onto teachers and each element in the domain (Students) is mapped to only one element in the range (Teachers).

Additional Readings

Stoll's book *Set Theory and Logic* (Dover edition, 1979) is an inexpensive yet fairly comprehensive and well-written introduction to set theory.

Other books that introduce set theory as well as functions include those on mathematical analysis. Bear's book *An Introduction to Mathematical Analysis* (Academic Press, 1997) is a truly simple and easy to understand introduction. Sprecher's book *Elements of Real Analysis* (Dover edition, 1987) is a more comprehensive book that includes the topics of sets and functions. Kolmogorov and Fomin's book *Introductory Real Analysis* (Dover edition, 1975) is another more comprehensive book that includes the topics of sets and functions.

Section II

Measure and Probability

3

Measure Theory

Suppose we have a set S = {Fred, Lisa, Sue} and we wish to assign numbers to various members of S. If we were only interested in the individual elements of S, we might assign a function from S to some set of numbers, perhaps representing the individual weight of three individuals referred to by the names *Fred*, *Lisa*, and *Sue*. However, suppose we also wanted to assign numbers to subsets of S; perhaps we are interested in the combined weight of Fred and Lisa. In this case, we are out of luck: a function assigns an element in its range to each element of its domain and cannot make an assignment to something that is not in its domain, including subsets of the domain elements themselves. Consequently, to achieve this goal we need to specify a domain for our number assignment that contains as an element any subset of S to which we wish to assign a number. If we denote the set of subsets of S to which we wish to assign numbers as A = {{Fred}, {Lisa}, {Sue}, {Fred, Lisa}, {Fred, Sue}, {Lisa, Sue}}, then the pair (S, A) is a way to denote a domain A for number assignment reflecting information in a set S we wish to represent: specifically, we wish to define a function on the subsets A of the set S. Regarding A, defined as above in terms of Fred, Lisa, and Sue, we can assign a number to each individual as we did before, but also to each pair of individuals, since A contains the sets {Fred, Lisa}, {Fred, Sue}, and {Lisa, Sue}. However, in this example we could not assign a number to the whole group because the set {Fred, Lisa, Sue} is not in A as we have defined it above. If we wish to assign a number to the whole group, we need to include the set {Fred, Lisa, Sue} in A as well: A = {{Fred}, {Lisa}, {Sue}, {Fred, Lisa}, {Fred, Sue}, {Lisa, Sue}, {Fred, Lisa, Sue}}.

Measurable Spaces

Formally, to properly operate as we will require, the set A in the preceding paragraph must be more structured than indicated by the preceding example. This is done by defining what is called an *algebra* as follows.

An *algebra* A defined on a set S is a set of subsets of S such that the following three conditions hold:

1. $S \in A$
2. If $A \in A$, then $\overline{A} \in A$
3. If $A \in A$ and $B \in A$, then $A \cup B \in A$

Here, $\overline{A}$ is defined with S as its universal set; in other words $\overline{A}$ is actually the relative complement of A with respect to S (i.e., $\overline{A} = S - A$). The three conditions imply that A is an algebra of S if A includes S itself, and it is closed under complementation and union. Note that this implies $\emptyset \in A$ because $S \in A$, and therefore $\overline{S} \in A$, but $\overline{S} = \emptyset$. Also, by this definition an algebra is closed under intersection and relative differences as well. This is the case because both intersection and relative differences can be expressed as equal to sets defined in terms of union and complementation alone: $A \cap B \approx \overline{(\overline{A} \cup \overline{B})}$ and $A - B \approx \overline{(\overline{A} \cup B)}$. You can show these equivalencies to yourself by the careful use of Venn diagrams.

Unfortunately, an algebra is not a sufficiently general structure to meet all of our needs; to meet our needs, we must modify this definition a bit to get what is called a *sigma-algebra* (also commonly denoted as a σ-algebra), defined as follows. The set A is a *sigma-algebra* of S if

1. $S \in A$
2. If $A \in A$, then $\overline{A} \in A$
3. If $\{A_n : n \in \mathbb{N}\}$ is a sequence of sets in A, then $\bigcup_n A_n \in A$

This is to say that A is a sigma-algebra of S if it includes S, and it is closed under complementation and countable unions. The notion of a countable union in Part (c) above is simply that the union of a sequence of sets in A is also in A. The detailed reasoning behind our move from unions to countable unions, and thereby from algebras to sigma-algebras, is beyond the scope of this book. Nonetheless, it turns out that sigma-algebras are sufficient for our use to represent finite, countable, and continuous sets.

For a set S and a corresponding sigma-algebra A, the pair (S, A) is called a *measurable space*. This pair of sets indicates that a collection of subsets of the set S has been identified to which we can assign numbers. Given the structure of a sigma-algebra, if we include a subset A of S in our sigma-algebra A, then not only is A measurable by virtue of being an element of A and thereby part of a domain on which we can define a function, but consequently $\overline{A}$ is also measurable. If we also include a subset B in A, then not only are B and $\overline{B}$ measurable but so is $A \cup B$, $A \cap B$, $A - B$, and $B - A$. Note that the smallest measurable space of a set S is $(S, \{S, \emptyset\})$; it is easy to verify that all conditions for a sigma-algebra are met by the set $\{S, \emptyset\}$.

For a set C, representing any collection of subsets of S, let $\sigma(C)$ denote the intersection of all sigma-algebras that contain C. The following two results hold for $\sigma(C)$: (1) $\sigma(C)$ is a sigma-algebra, and (2) $\sigma(C)$ is the smallest sigma-algebra containing C. The set $\sigma(C)$ is the minimal sigma-algebra generated by C. The details of these consequences are not of concern in this book; what *is* important is that they imply we can construct a measurable space for a set S by identifying a set C of subsets that we wish to measure and defining an associated sigma-algebra as $\mathcal{A} = \sigma(C)$. It is often easy to proceed by recursively taking all the complements, unions, intersections, and relative differences of the collection C, and consequent sets, to get $\sigma(C)$. It becomes even easier to use $\sigma(C)$ if we identify C with some partition of interest. This is a natural case for applied researchers because we often think in terms of qualities that partition sets of objects. For example, consider the quality of *age* (in whole units of years) associated with people in some population: this quality can be used to partition the population into sets of people having the same age—those 1 year old, 2 years old, 3 years old, and so on. The sigma-algebra generated by this partition allows numbers to be assigned to any age subgroups of the population, since it includes the union of any age subsets.

Note that I don't need the full partition to generate the sigma-algebra; I can leave out one of the sets. In other words, if a partition comprises K sets, I can, but don't need to, generate the sub-sigma-algebra using only $K - 1$ of the sets.

Measures and Measure Spaces

In the preceding section, we defined measurable spaces as mathematical structures that allow us to assign numbers to subsets of a given set, but we have yet to define a useful assignment of such numbers. This section provides an important type of assignment, one that will underlie probability theory as well as a general calculus of integration. For a measurable space $(S, \mathcal{A})$, a function μ is called a *measure* if it has domain $\mathcal{A}$ and range contained in the extended real line such that

1. $\mu(A) \geq 0$, for all $A \in \mathcal{A}$
2. $\mu(\varnothing) = 0$
3. For any disjoint collection of sets $\{A_n : n \in \mathbb{N}\}$ in $\mathcal{A}$

$$\mu\left(\bigcup_n A_n\right) = \sum_n \mu(A_n)$$

By this definition, it is clear that a measure assigns each set in a sigma-algebra to a number in the nonnegative real line. It must assign the empty set to 0 (and may assign other sets to 0 as well), and it must assign the union of disjoint sets to the same number as the sum of what it assigns to each of the individual sets.

If $(S, \mathcal{A})$ is a measurable space and μ is a measure with domain $\mathcal{A}$, then $(S, \mathcal{A}, \mu)$ is called a *measure space*. Note that there are many ways, true to the definition of a measure, to map a given $\mathcal{A}$ to the nonnegative real line; therefore, many measure spaces can be associated with a single measurable space.

A measure space $(S, \mathcal{A}, \mu)$ has three useful properties that we are concerned with at the moment. For sets A and B and the sequence of sets $\{A_j : j \in \mathbb{N}\}$, all of which are members of $\mathcal{A}$,

1. If $A \subseteq B$, then $\mu(A) \le \mu(B)$
2. $\mu(A \cup B) + \mu(A \cap B) = \mu(A) + \mu(B)$
3. $\mu\left(\bigcup_n A_n\right) \le \sum_n \mu(A_n)$

Property a is evident if we define a set C to be the relative difference of set B with respect to set A (i.e., $C \approx B - A$). Consequently, set B is the union of the disjoint sets A and C (i.e., $B \approx A \cup C$), and the measure of B is equal to the sum of the measures of A and C [i.e., $\mu(B) = \mu(A) + \mu(C)$] from which the property is directly observed due to the fact that μ is a nonnegative function. Note that equality holds in property a if C is the empty set or any set with measure equal to 0.

Property b is evident if we note that the union of A and B can be expressed as the union of three disjoint sets—specifically, two relative differences and an intersection:

$$(A \cup B) \approx (A - B) \cup (B - A) \cup (A \cap B) \tag{3.1}$$

Being disjoint sets, the measure of the union of A and B is therefore the sum of the measure of each component:

$$\mu(A \cup B) = \mu(A - B) + \mu(B - A) + \mu(A \cap B) \tag{3.2}$$

Consequently, by adding $\mu(A \cap B)$ to both sides of Equation 3.2, the right-hand side can be expressed as follows:

$$\underbrace{\mu(A - B) + \mu(A \cap B)}_{\text{Part 1}} + \underbrace{\mu(B - A) + \mu(A \cap B)}_{\text{Part 2}} \tag{3.3}$$

However, Part 1 of Equation 3.3 is the measure of two disjoint sets that compose the set A and is therefore $\mu(A)$, and Part 2 is the measure of two disjoint sets that compose set B and is therefore $\mu(B)$. Property b immediately follows.

Property c is evident from extending property b over a sequence of sets and dropping the terms representing the intersections, thereby generating the inequality.

Example 3.1

Let S be the set of N people in a given room: $S = \{s_1, s_2, \ldots, s_N\}$. Let C denote the set of subsets containing each individual: $C = \{\{s_1\}, \{s_2\}, \ldots, \{s_N\}\}$. Then we can define a sigma-algebra on S as the sigma-algebra generated by C, which turns out to be the power set of S: $A = \sigma(C) = \wp(S)$.

Is (S, A) in Example 3.1 a measurable space? Because $\sigma(C)$ is by definition a sigma-algebra on S, then the answer is "yes" by construction—OK, that was too easy. It is more interesting to consider how we can see that (S, A) is a measurable space. Because A is the power set of S, it contains all subsets of S. The power set must therefore contain all complements, because the complement of a subset of S is also a subset of S (i.e., all the elements of S that are not in the targeted subset). The power set also contains all intersections of subsets of S because a set of elements of S that are members of any number of other subsets are still members of S and therefore compose a subset of S. Moreover, because S itself is a subset of S, and the complement of S is $\emptyset$, then both S and $\emptyset$ are also members of the power set. It is evident that all conditions for (S, A) to be a measurable space are met.

Consider a function that counts the elements of a set, for example, $\mu(\{s_i\}) = 1$ for all $i \in \{1, \ldots, N\}$. Is μ a measure on (S, A) in Example 3.1? Because a set is either empty or contains a positive number of elements, then $\mu(A) \geq 0$, for all $A \in A$. Moreover, the number of elements in the empty set is 0, so $\mu(\emptyset) = 0$. Suppose A is a nonempty subset of S that is a member of A; such a subset contains a group of individuals in the classroom and is therefore the union of some sets in C, which are disjoint. The measure of A is the sum of the measures of the corresponding sets in C, which is $\mu(\cup_n A_n) = \sum_n \mu(A_n)$. The function μ is therefore a measure on (S, A).

Example 3.2

Let S be the set of N people in a given room: $S = \{s_1, s_2, \ldots, s_N\}$. Let C contain the subset of women: $C = \{\{\text{women}\}\}$, where I use $\{\text{women}\}$ as shorthand for the subset of women in S. Then $A = \sigma(C) = \{S, \emptyset, \{\text{women}\}, \{\text{men}\}\}$ is a sigma-algebra on S. The pair (S, A) is a measurable space. Specify a measure μ on (S, A) such that $\mu(\{\text{women}\}) = |\{\text{women}\}| / |S|$ and $\mu(\{\text{men}\}) = |\{\text{men}\}| / |S|$, for which the parallel lines $|\cdot|$ denote the number of members of a set. Therefore $\mu(A)$ is the proportional size of A with respect to S. Because μ is a measure, then considering the union of the disjoint sets of women and men, $\mu(S) = 1$; and because $\mu(S \cup \emptyset) = \mu(S) + \mu(\emptyset)$ but the union of S and $\emptyset$ is S, then $\mu(S) = \mu(S) + \mu(\emptyset)$, which implies $\mu(\emptyset) = 0$, as we expect.

Note that in Example 3.2 we did not specify a measure that was able to assign numbers to individuals in the room, or any subsets other than those in A. Consequently, in this case the measure space (S, A, μ) is fairly limited in representing information about S; specifically, our analysis is restricted to $A \approx \{S, \emptyset, \{\text{women}\}, \{\text{men}\}\}$.

Measurable Functions

If f denotes a one-to-one function, it is common to denote its inverse (the function that maps the range of f back to its domain) as f^{-1}: we do *not* use this notational convention here! Instead, we define f^{-1} as $f^{-1}(B) = \{a: f(a) \in B\}$. In other words, f^{-1} identifies the subset of the domain for f that f maps to a specified subset of its range. Consequently, f^{-1} takes subsets of the range for f as its argument and returns subsets of the domain of f. This inverse *set function* f^{-1}, therefore, does not require the function f to be one-to-one. Suppose for example that f maps its whole domain to a single constant; f^{-1} would then map that constant back to the set that constitutes the full domain of f.

Let $(S, \mathcal{A})$ and $(T, \mathcal{F})$ be two measurable spaces. If f is a function from S to T such that $f^{-1}(B) \in \mathcal{A}$ for all $B \in \mathcal{F}$, then f is called a *measurable function* because we can define a measure on $(T, \mathcal{F})$ in terms of a measure on $(S, \mathcal{A})$, as shown below. Note that the domain of f^{-1} includes the sigma-algebra $\mathcal{F}$ associated with T, the range of f. The key to the definition of a measurable function is that for all sets in $\mathcal{F}$ (the sigma-algebra we specified as associated with the range of f) there is a corresponding set in $\mathcal{A}$ (the sigma-algebra we specified as associated with the domain of f) to which f^{-1} maps. Note, however, that it is not required that every set of S included in $\mathcal{A}$ be mapped by f^{-1} from a set in $\mathcal{F}$; only the reverse is required.

We can define a measure on $(T, \mathcal{F})$ in terms of a measure on $(S, \mathcal{A})$ as follows. Let $(S, \mathcal{A}, \mu)$ be a measure space, $(T, \mathcal{F})$ be another measurable space, and f be a measurable function from S to T. Then, we can define a function, say μ^*, on the sigma-algebra $\mathcal{F}$ as $\mu^*(B) = \mu(f^{-1}(B))$, yielding $(T, \mathcal{F}, \mu^*)$ as another measure space. The measure μ^* assigns to each set $B \in \mathcal{F}$ the number that μ assigns to the corresponding set $A = f^{-1}(B)$ in $\mathcal{A}$. So, if for each set B in $\mathcal{F}$, the inverse set function $f^{-1}(B)$ corresponds to a set A in $\mathcal{A}$, then there is the following flow between the two measure spaces:

$$(S \quad \mathcal{A} \quad \mu)$$
$$\downarrow \quad \uparrow \quad \downarrow$$
$$(T \quad \mathcal{F} \quad \mu^*)$$

indicating how the measure space $(T, \mathcal{F}, \mu^*)$ is generated from $(S, \mathcal{A}, \mu)$. The leftmost arrow represents f, the middle arrow represents the corresponding inverse set function f^{-1} as applied to events in $\mathcal{F}$, and the rightmost arrow represents the resulting specification of μ^* as an extension of μ [i.e., for each $B \in \mathcal{F}$, $\mu^*(B) = \mu(f^{-1}(B))$].

Integration

Measure theory provides a definition and understanding of integration that is more general than what is provided in typical undergraduate introductory courses. For the purpose of this book, however, we need not expound on the details; understanding basic concepts will suffice.

First, let's develop a simple understanding from the basic introductory courses. The integral of some function f with respect to its domain X is taken to be the sum (denoted by the symbol $\int$) across the domain X of the function evaluated at each element x in the domain multiplied by some small length at x (denoted as dx). Essentially, an integral is a sum of products, commonly denoted as:

$$\int f(x) \cdot dx \qquad (3.4)$$

This basic "sum of products" notion holds in the more general measure-theoretic definitions of integration as well, in which we change "length" to "measure." Before we provide that definition, however, it would be helpful to start with the basic concepts of measure spaces and measurable functions and work up from there.

Suppose we have a measure space $(S, \mathcal{A}, \mu)$ with f a nonnegative real-valued function of S measurable $\mathcal{A}$; moreover, suppose $\mathcal{A}$ is the power set of S. The function f assigns numbers to its domain S, and μ assigns numbers to its domain, each element of $\mathcal{A}$, including the singletons that comprise individual elements of S. Suppose that for each $s \in S$ and corresponding $\{s\} \in \mathcal{A}$, I took the product $f(s) \cdot \mu(\{s\})$ and then summed across all s, what would I have? I'd have a sum of products, which indeed, by the definition provided below, would be the integral of f with respect to μ. Albeit a bit more complicated, the measure-theoretic definition of integral will entail summing products of a measurable function with a corresponding measure. The greater complexity comes from the fact that $\mathcal{A}$ need not be the power set of S (or a continuous-space analog such as the Borel sigma-algebra).

From a measure-theoretic perspective, one useful definition of an integral for a nonnegative measurable function f relative to the measure of a measure space $(S, \mathcal{A}, \mu)$ is

$$\int f \cdot d\mu = \sup_{\pi \in \Pi} \sum_{A \in \pi} \left[\left(\inf_{w \in A} f(w) \right) \cdot \mu(A) \right] \qquad (3.5)$$

where Π denotes the set of all partitions of the set S in the sigma-algebra $\mathcal{A}$, inf (the infimum) denotes the greatest lower bound, and sup (the supremum) denotes the least upper bound. For finite sets, the inf and sup correspond to the minimum and maximum, respectively.

The conventional notation $\int f \cdot d\mu$, which has a familiar look to those versed in undergraduate calculus, can be a bit confusing, since the definition of $d\mu$ as a differential is a bit unclear when applied to an arbitrary measure space. Looking at the preceding definition, which contains the sum of the product of function values and measure values across partitions without reference to differentials, it might be better to denote the integral simply as $\int f \cdot \mu$. Later in the text, I will periodically use the second notation; however, for now I will use the more common first notation. Both denote the same thing.

To simplify the following discussion, it will be helpful to denote the summation as:

$$G(\pi) = \sum_{A \in \pi} \left[\left(\inf_{w \in A} f(w) \right) \cdot \mu(A) \right] \tag{3.6}$$

and therefore

$$\int f \cdot d\mu = \sup_{\pi \in \Pi} G(\pi) \tag{3.7}$$

Note, however, that there are other definitions of integrals, many of which lead to essentially the same results; others do not. For example, one could switch the sup and the inf to obtain another definition, which for smooth continuous functions would provide the same answer as the above definition when integrated with respect to a Borel sigma-algebra. For our purpose, we will use the above definition.

To see how integration works by our definition, let's break it down into steps. First, consider any particular partition π of the set S, the elements of which are contained in our sigma-algebra $\mathcal{A}$ (and is thereby subject to the measure μ). Being a partition, π comprises sets that are mutually exclusive and together make up the set S. Second, for each set A in the partition π, find the element of A that has the smallest value of the function f. Third, multiply this least value of f in A by the measure μ of the set A. Fourth, sum this product across all sets in the partition π. Fifth, repeat the second through fourth steps for all other partitions of S in the sigma-algebra $\mathcal{A}$. The integral of f with respect to μ is the largest of these sums.

Example 3.3

Consider S to be the pattern of dots on the six faces of a die, which I will denote as [1], [2], [3], ... , [6], in which [x] means the face with x number of dots. Let $\mathcal{A}$ be $\{S, \emptyset, \{[1], [2], [3]\}, \{[4], [5], [6]\}\}$. Let f assign the number of dots represented by each element of S (i.e., $f([x]) = x$), and let μ assign the total number of dots represented by each set in $\mathcal{A}$ (i.e., $\mu(\{[x], [y], [z]\}) = x + y + z$). First, note that there is only one partition of S represented in $\mathcal{A}$, specifically $\{\{[1], [2], [3]\}, \{[4], [5], [6]\}\}$; consequently identifying the integral of f with respect to μ does not require searching over a whole set of partitions in this example. Now, because $f([x]) = x$, the smallest f associated with $\{[1], [2], [3]\}$ is $f([1]) = 1$, and the smallest f associated with $\{[4], [5], [6]\}$ is $f([4]) = 4$.

Moreover, $\mu(\{[1], [2], [3]\}) = 1 + 2 + 3 = 6$ and $\mu(\{[4], [5], [6]\}) = 4 + 5 + 6 = 15$. Therefore, the integral of f with respect to μ is:

$$
\begin{aligned}
\int f \cdot d\mu &= \sup_{\pi \in \Pi} \sum_{A \in \pi} \left[\left(\inf_{w \in A} f(w) \right) \cdot \mu(A) \right] \\
&= f([1]) \cdot \mu(\{[1], [2], [3]\}) + f([4]) \cdot \mu(\{[4], [5], [6]\}) \qquad (3.8) \\
&= 1 \cdot 6 + 4 \cdot 15 \\
&= 66
\end{aligned}
$$

Example 3.4

Suppose in the preceding example $\mathcal{A}$ is instead defined as the power set of S, and f and μ are defined as above. There are now many partitions of S in $\mathcal{A}$. It turns out, however, that the most granular partition has the largest sum and is thereby used to calculate the integral, which in this case is

$$
\begin{aligned}
\int f \cdot d\mu &= \sup_{\pi \in \Pi} \sum_{A \in \pi} \left[\left(\inf_{w \in A} f(w) \right) \cdot \mu(A) \right] \\
&= \sum_{x \in \{1, \ldots, 6\}} f([x]) \cdot \mu(\{[x]\}) \qquad (3.9) \\
&= \sum_{x \in \{1, \ldots, 6\}} x^2 \\
&= 91
\end{aligned}
$$

Integrals are primarily driven by the finest measurable partitions [i.e., the finest partitions produce the largest sums $G(\pi)$]. This is the case because for two partitions π_0 and π_1, if π_1 is a refinement of π_0, then $G(\pi_1) \geq G(\pi_0)$. This can be shown by considering an arbitrary set A in π_0 and the corresponding sets A_i in π_1 that make up A. Because the sum of the measures of the π_1 sets is equal to the measure of the π_0 set under consideration, one can determine that the $(\inf_{w \in A} f(w)) \cdot \mu(A)$ term associated with $G(\pi_0)$ cannot be larger than $\sum_{A_i \in A} [(\inf_{w \in A_i} f(w)) \cdot \mu(A_i)]$ associated with $G(\pi_1)$. This is due to the fact that $(\inf_{w \in A} f(w)) \cdot \mu(A) = \sum_{A_i \in A} [(\inf_{w \in A} f(w)) \cdot \mu(A_i)]$, where the inf term is constant across the summation that is the smallest value: the sum of this smallest constant cannot be any larger than a sum that may include larger values but necessarily does not include smaller values.

Unfortunately, if π_1 and π_2 are different refinements of π_0, then even though $G(\pi_1) \geq G(\pi_0)$ and $G(\pi_2) \geq G(\pi_0)$, we do not know *a priori* whether $G(\pi_1) \geq G(\pi_2)$ or $G(\pi_2) \geq G(\pi_1)$. Similarly, we cannot judge *a priori* between two different refinements of two different partitions. In other words, we have yet to isolate a single partition π^* such that $\int f \cdot d\mu = G(\pi^*)$. However, it turns out that if we have a partition in $\mathcal{A}$ that is a refinement of all other partitions in $\mathcal{A}$, this partition is just such a π^*. This is evident because as a refinement of every partition in $\mathcal{A}$, $G(\pi^*) \geq G(\pi)$ for all π in $\mathcal{A}$, and there exists no further refinement of π^* in $\mathcal{A}$ to achieve an even greater G value.

For the measure space $(S, \mathcal{A}, \mu)$ where S is a countable set and $\mathcal{A}$ is the power set of S, the partition made up of the singletons comprising each member of S is a partition π^* such that $\int f \cdot d\mu = G(\pi^*)$. If conversely S is a continuous set (e.g., some interval on the real line), then the Borel sigma-algebra contains the necessary analog to π^* to allow integration of continuous variables (see more advanced texts in measure theory for details on Borel sets and their application).

As an aside, note that by this definition, integration automatically represents summation across discrete sets: consequently, one technically does not need to use separate notation for discrete and continuous variables—although in most cases people do so anyway, using the summation sign (Σ) for discrete sets.

At the beginning of this section, I framed the definition of an integral in terms of a nonnegative function. We can easily extend the preceding results to functions on the whole real line by partitioning the domain of the function into a set that is mapped to the nonnegative real line and another set that is mapped to the negative real line. Next, evaluate the integral of each part separately, except use the absolute value of the function for the region mapping to the negative real line. Then subtract the integral of the negative part from the integral of the positive part to achieve the full integral of the function.

The goal of this section is not to exhaust the reader's patience with excess detail; the typical undergraduate-level understanding of integration will suffice in operationalizing probability theory for most applied purposes (although the current definition is helpful when operationalizing Monte Carlo techniques for numeric integration). Rather, the purpose of this section is to set the stage for understanding a particular type of function in terms of measure theory: the density function.

Suppose we have a measure space $(S, \mathcal{A}, \mu)$ and a nonnegative measurable function f on S; the following integral defines another measure v on $(S, \mathcal{A})$:

$$v(A) = \int_A f \cdot d\mu, \quad \text{for all } A \in \mathcal{A} \tag{3.10}$$

In this case, the function f is a *density* function, and more specifically f is the density for the measure v with respect to μ.

Although density functions are not restricted to continuous measure spaces because our definition of integral encompasses finite, countable, and continuous variables, they are particularly important when characterizing continuous spaces, that is to say when we are concerned with continuous variables. When considering probability theory, we may start with the measure v and identify a density for that measure as a nonnegative function f that makes the preceding relation true, or we may start with a density f and identify the corresponding measure v.

Additional Readings

Because measures and probability are so intimately related, additional readings for this section are essentially the same as those cited in the section "Additional Readings" at the end of Chapter 4, "Probability." See that section for additional readings appropriate to this chapter.

4

Probability

If $(\Omega, \mathcal{A}, P)$ is a measure space as defined in the previous chapter and the measure P assigns 1 to the set Ω (i.e., $P(\Omega) = 1$), then P is called a *probability measure* or sometimes just a *probability*, and the triple $(\Omega, \mathcal{A}, P)$ is then called a *probability space*. Because $P(\Omega) = 1$, P assigns a number no greater than 1 to any set in the sigma-algebra $\mathcal{A}$, and the sum of probabilities assigned by P to the sets of a partition must sum to 1 across the partition, which is just what we expect from a probability. Note that for any measure space $(\Omega, \mathcal{A}, \mu)$ with $\mu(\Omega)$ finite and nonzero, a new measure can be defined by dividing μ's assignment for any set by the number that μ assigns to Ω. That is, $P(A) = \mu(A)/\mu(\Omega)$ for all sets A in the sigma-algebra $\mathcal{A}$ is a probability measure on $(\Omega, \mathcal{A})$.

Example 4.1

Let Ω be the set of possible outcomes of the flip of a coin specified as $\Omega = \{heads, tails\}$. Define $\mathcal{C} = \{\{heads\}\}$, and therefore $\mathcal{A} = \sigma(\mathcal{C}) = \{\Omega, \varnothing, \{heads\}, \{tails\}\}$. We can define an infinite number of probability measures on the measurable space $(\Omega, \mathcal{A})$; some are shown in Table 4.1.

Note that $(\Omega, \mathcal{A}, P)$ is a probability space by virtue of its mathematical properties alone. What $(\Omega, \mathcal{A}, P)$ is taken to mean is an entirely separate, and nonmathematical, question. The meaning of $(\Omega, \mathcal{A}, P)$ comes from our interpretation of it as a model of something in which we are interested. Researchers commonly use probability spaces to model uncertainty associated with data generating processes and to model subjective beliefs. Other interpretations exist as well. See the section "Additional Readings" at the end of Chapter 5 for references to books that discuss various interpretations of probability.

A *data generating process* is the mechanism by which observational units are obtained and measurements are made. The data generating process is typically taken to be objective. Uncertainty from a data generating process is created if the process has the potential of producing different observational units or measurements. For example, one data generating process is an equal probability random sample from a fixed population. The outcome of the process is a particular member of the population; because the process is a random sampling mechanism, it can potentially produce any one of the members. Hence, we are uncertain, *a priori*, about the result of the data

TABLE 4.1

Probabilities for Coin Flip Outcome Events

Probabilities	Ω	$\varnothing$	{heads}	{tails}
			\multicolumn{2}{c}{$\mathcal{A}$}	
P_1	1	0	0.50	0.50
P_2	1	0	0.75	0.25
P_3	1	0	0.02	0.98

generating process. We use probability spaces to model this process and uncertainty.

Another common use of probability spaces is to model *subjective uncertainty*: what a person believes. For example, we may model how confident a person is that a given outcome will occur, or we may model a person's certainty in a particular value of an underlying parameter (e.g., the average height among US citizens).

Example 4.2

Suppose we wish to model the probability of outcomes associated with a flip of a coin as in Example 4.1. If the coin is fair, P_1 would represent a model of the data generating process. Of course, P_1 can also represent a model of a person who believes there is a 50% chance of heads, or who is indifferent between whether heads or tails will occur. The remaining probabilities shown in Table 4.1 could also be used for either a data generating process, if the process was not fair, or subjective uncertainty, if people believe the process was not fair.

For a probability space $(\Omega, \mathcal{A}, P)$ modeling a data generating process, the set Ω is often called the *outcome set* and represents the set of possible outcomes from the data generating process. The set $\mathcal{A}$ is commonly called the *event set* and represents the events to which we wish to assign probabilities. Note that events are sets of outcomes, whereas outcomes are particular possibilities: If I role a numbered six-sided die, I might get the event of an even number, which is a number in the set {2, 4, 6}, or perhaps I will get an event of less than 5, which is a number in the set {1, 2, 3, 4}, or maybe the event of a 2, which is a number in the set {2}. However, the actual outcome is a specific number from 1 to 6. The measure P is the probability measure representing the uncertainty about the occurrence of events.

You need to define P as a model of something useful for your purpose! This may be a model to represent a source of uncertainty, or it may be a model to represent a normalized frequency (no uncertainty implied, as in a population model), or yet some other interpretation.

Example 4.3

If I were a javelin thrower, then Ω might be the possible outcomes of a throw—a set of distances. If I wanted to model the probability of throwing at least 100 feet, then one of the subsets in $\mathcal{A}$ must represent the event of throwing a distance of 100 feet or more.

In many types of research, it is common to use the following specifications:

Ω is some population of interest (e.g., a population of people or hospitals).

$\mathcal{A}$ is the power set of Ω, allowing for probabilities to be assigned to any subset.

P is based on the sampling probabilities of the individuals w in the population, $w \in \Omega$. In other words, there is a sampling probability for each w in Ω reflecting the data generating process, and the $P(A)$ for all $A \in \mathcal{A}$ is derived additively across unions of the singletons $\{w\}$ for all $\{w\}$ in A.

Such a specification for $(\Omega, \mathcal{A}, P)$ is typical for representing an objective data generating process. Users of such models are often called *frequentists* because they are assumed to be interpreting P as long-run frequencies associated with the data generating process. There are other useful objective interpretations for P, the propensity interpretation of probabilities being one example, but the frequentist label has stuck. Because it seems to me that modeling the data generating process as an objective mechanism remains the most common use for specifying a probability space in applied research, this book will take the frequentist perspective for most examples. However, the general principles apply to any interpretation of probability because the mathematics do not change by virtue of the interpretation.

Conditional Probabilities and Independence

Consider a probability space $(\Omega, \mathcal{A}, P)$. If sets A and B are elements of the sigma-algebra $\mathcal{A}$ and are thereby available for number assignment by probability measure P, then, as required by our definition of a measure space, their intersection is also measurable by P. So we have a probability associated with an event of both A and B occurring, that is, $P(A \cap B)$. If the set B is such that $P(B) \neq 0$, then we can define the following ratio, which is itself a probability measure on $(\Omega, \mathcal{A})$. Let $P_B(A)$ denote this ratio:

$$P_B(A) = \frac{P(A \cap B)}{P(B)} \tag{4.1}$$

for all $A \in \mathcal{A}$ and any set $B \in \mathcal{A}$ with $P(B) \neq 0$.

Challenge 4.1

Describe a probability space for 10 people who are randomly assigned to a treatment group or a control group with equal probability.

The triple $(\Omega, \mathcal{A}, P_B)$ is a probability space, and P_B as defined in Equation 4.1 is the *conditional probability* given B, also denoted as $P(A \mid B)$. Note that conditional probabilities are only defined relative to certain subsets of Ω, specifically sets that are members of $\mathcal{A}$ and that have a nonzero probability. Consequently, if $\mathcal{A}$ is defined in a coarse manner (i.e. defined in terms of only a few subsets of all the possible subsets of Ω), then only a few conditional probabilities can be considered.

Example 4.4

For $(\Omega, \mathcal{A}, P)$ specified such that Ω is the set of individuals in a given room and $\mathcal{A} = \{\Omega, \varnothing, \{\text{women}\}, \{\text{men}\}\}$, then the conditional probability $P_{\{women\}}(A)$ is equal to either 1 or 0 depending on the A in $\mathcal{A}$. Nothing more interesting can be discerned. If instead we define $\mathcal{A} = \sigma(\{\{\text{women}\}, \{\text{left handed}\}\})$, then $P_{\{women\}}(A)$ would assign a probability of being left-handed among women and a probability of right-handed among women, which would sum to 1 (assuming for simplicity there are only left- and right-handed people).

So far, by defining a conditional probability, we have not changed the measurable space but only changed the probability measure itself. We can, however, define a new probability space with a new measurable space as well. This can be achieved by defining a *trace* of the sigma-algebra $\mathcal{A}$ on the set $B \in \mathcal{A}$ as the set $\mathcal{A}_B = \{B \cap A : A \in \mathcal{A}\}$. This is the set comprising the intersections between B and all sets in $\mathcal{A}$. It can be shown that $\mathcal{A}_B$ is itself a sigma-algebra and $\mathcal{A}_B \subseteq \mathcal{A}$. The triple $(B, \mathcal{A}_B, p)$, with $p(C) = P_B(A)$, for all $C \in \mathcal{A}_B$ such that $C = A \cap B$, is a probability space identical to $(\Omega, \mathcal{A}, P_B)$ on all sets with $P_B(A) \neq 0$. In other words, a conditional probability can be represented by considering the conditioning event B as an outcome set and applying an appropriately normalized probability to the trace of $\mathcal{A}$ on B. In this case, the resulting probability p is not a conditional probability defined on $(\Omega, \mathcal{A})$ but rather a probability on the measurable space $(B, \mathcal{A}_B)$, yet it contains essentially the same information regarding $\mathcal{A}_B$ as the conditional probability $P_B(A)$.

For probability spaces $(\Omega, \mathcal{A}, P)$ and $(\Omega, \mathcal{A}, P_B)$, events A and B are *independent* elements of $\mathcal{A}$ if $P(A) = P_B(A)$. This relationship for independent events A and B requires that B have nonzero probability; it follows however from the more general definition of independence, which is $P(A \cap B) = P(A) \cdot P(B)$. Both imply, for B having nonzero probability, that the probability of an event A is equal to the probability of the intersection of A and B relative to the probability of B. In terms of a Venn diagram as

FIGURE 4.1

Independent events: events represented as the sets *A* and *B* are independent if the proportion of Ω occupied by the area of *A* (the combined light and dark shaded areas) is the same as the proportion of *B* occupied by the intersection of *A* and *B* (the dark shaded area); if not, *A* and *B* are dependent.

shown in Figure 4.1, this means the proportion of the overall area occupied by set *A* is the same as the proportion of set *B* that set *A* overlaps. That is, the proportion of Ω that set *A* covers is the same as the proportion of *B* that set *A* covers.

Product Spaces

Focusing on the data generating process as an example, we consider $(\Omega, \mathcal{A}, P)$ as a model of the uncertainty for the process that selects an outcome from Ω and thereby generates events in $\mathcal{A}$. Outcomes of the process, however, may be complex. For example, it might be that an experiment selects both a subject and an intervention to apply to that subject. The pair (subject, intervention) may vary in terms of which subject gets selected and which intervention gets applied. Another example is a data generating process that selects *N* subjects (i.e., gets a sample size of *N*). These and other complex outcomes can be factored into simpler components.

Example 4.5

Suppose we randomly select a subject from a population and give that person a randomly selected intervention from a set of possible interventions. The outcome would be a subject and intervention pair. Each component of the (subject, intervention) outcome may be represented by its own outcome space. Perhaps Φ is the set of possible subjects (e.g., the population from which a subject is to be drawn) and Θ is the set of possible interventions. It is reasonable to assume we can define corresponding measurable spaces $(\Phi, \mathcal{F})$ and $(\Theta, \mathcal{Q})$. From these individual measurable spaces, we should be able to construct a more complex measurable space $(\Omega, \mathcal{A})$. Indeed we can.

TABLE 4.2

Simple Cartesian Product

		Θ		
		θ_1	θ_2	θ_3
Φ	ϕ_1	(ϕ_1, θ_1)	(ϕ_1, θ_2)	(ϕ_1, θ_3)
	ϕ_2	(ϕ_2, θ_1)	(ϕ_2, θ_3)	(ϕ_2, θ_3)

Let $(\Phi, \mathcal{F})$ and $(\Theta, \mathcal{Q})$ be any two measurable spaces. The *Cartesian product,* or simply the *product,* of the outcome sets Φ and Θ is denoted as $\Phi \times \Theta$ and is defined as the set of all outcome pairs (w, x) such that $w \in \Phi$ and $x \in \Theta$: that is, $(w, x) \in \Phi \times \Theta$. If both Φ and Θ are finite, one could simply create a table for which the rows represent the members of Φ and the columns represent the members of Θ; the cells of the table represent the corresponding pairs (w, x). In this case, the table fills out the members of the product $\Phi \times \Theta$. Suppose $\Phi = \{\phi_1, \phi_2\}$ and $\Theta = \{\theta_1, \theta_2, \theta_3\}$, then $\Phi \times \Theta$ would contain as elements the six pairs of (ϕ_i, θ_j) shown in Table 4.2.

Because a sigma-algebra is a set (specifically, a set of sets), we can take the products of sigma-algebras as well. The product of sigma-algebras creates the set of all pairs comprising each set in one sigma-algebra with the sets in the other. The product of the two sigma-algebras $\mathcal{F}$ and $\mathcal{Q}$ can be used to generate a sigma-algebra on $\Phi \times \Theta$; the resulting sigma-algebra is denoted as $\mathcal{F} \otimes \mathcal{Q}$, such that $\mathcal{F} \otimes \mathcal{Q} = \sigma(\mathcal{F} \times \mathcal{Q})$. In other words, take all the defined events in $\mathcal{F}$, cross them with all the defined events in $\mathcal{Q}$, and take all complements and unions, according to the definition of a sigma-algebra, to produce a new sigma-algebra based on the product of the original two. Then $(\Phi \times \Theta, \mathcal{F} \otimes \mathcal{Q})$ is a measurable space corresponding to $(\Omega, \mathcal{A})$ in which $\Omega = \Phi \times \Theta$ and $\mathcal{A} = \mathcal{F} \otimes \mathcal{Q}$.

For the probability space $(\Phi \times \Theta, \mathcal{F} \otimes \mathcal{Q}, P)$, the probabilities defined as $P_\Phi(F) = P(F \times \Theta)$ for all $F \in \mathcal{F}$ and $P_\Theta(Q) = P(\Phi \times Q)$ for all $Q \in \mathcal{Q}$ are called the *marginal probabilities.* The probability $P_\Phi(F)$ is therefore the probability of an outcome in Φ that is in the set F and *any* element of Θ, whereas $P_\Theta(Q)$ is the probability of an outcome with any element of Φ and an element of Θ that is in the set Q. The marginal probabilities are associated with the probability spaces $(\Phi, \mathcal{F}, P_\Phi)$ and $(\Theta, \mathcal{Q}, P_\Theta)$, sometimes called *coordinate spaces,* which correspond to incomplete observation of the data generating process that produces $(w, x) \in \Phi \times \Theta$. However, in the former case we only observe outcome w, and in the latter case we only observe outcome x. The coordinate spaces are independent if $P(F \times Q) = P_\Phi(F) \cdot P_\Theta(Q)$ for all $F \in \mathcal{F}$ and $Q \in \mathcal{Q}$.

Note that the notation used here for marginal probabilities is the same as that used in the preceding discussion of conditional probabilities. This is purely for typographical convenience when writing the probability space associated with a conditional distribution; it is cleaner to write $(\Omega, \mathcal{A}, P_B)$ than say $(\Omega, \mathcal{A}, P(\cdot \,|\, B))$ for a conditional probability space. The context will make clear which probability is meant, and when both are used together the more common $P(A \,|\, B)$ notation will be used for conditional probabilities, as in the following paragraph.

A probability of an event in one coordinate given an event in another coordinate can also be defined on $(\Phi \times \Theta, \mathcal{F} \otimes \mathcal{Q}, P)$ using the definition of conditional probability above. $P_{\Phi \times Q}(F \times \Theta)$, also denoted as $P(F \times \Theta \,|\, \Phi \times Q)$, is the probability of an event F in the sigma-algebra $\mathcal{F}$, and the event of "any outcome" in the sigma-algebra $\mathcal{Q}$ given an event Q in the sigma-algebra $\mathcal{Q}$, and the event of "any outcome" in the sigma-algebra $\mathcal{F}$. If we were representing $\mathcal{F} \times \mathcal{Q}$ as a table, with the events in $\mathcal{F}$ as rows and the events in $\mathcal{Q}$ as columns, then $P_{\Phi \times Q}(F \times \Theta)$ would represent the probability of a row representing the event F given a column representing the event Q. It should be clear that because we are conditioning on a particular event in $\mathcal{Q}$, we can denote this probability as $P_{\Phi \times Q}(F)$ without confusion. The coordinate spaces are dependent if the conditional probability is not equal to the marginal probability across events.

Product spaces are more general than the two-dimensional ones discussed above. Any number of outcome sets can be components of a product space: for example, N individual outcome sets can compose an overall product set $\Phi_1 \times \Phi_2 \times \ldots \times \Phi_N$. If each is finite, then the product space can be thought of as an N-dimensional array in which each cell is a particular possible outcome such as $(\phi_1, \phi_2, \ldots, \phi_N)$.

Dependent Observations

Now we get to a question important to applied researchers that measure theory can help us sort out: when are observations dependent? Before answering, however, we need to better understand what it is that we are taking to be dependent or independent. When modeling a data generating process, the use of the term *observation* is unfortunate, as it leads one to think about a particular realization of the process: the actual result. However, our probability space is a model of the possible results from a data generating process, not an actual realization. Unfortunately, *observation* is the term in common use, so I will use it here. However, by *observation* what I really mean is "potential result of engaging a given data generating process"—that is, a specific process of observation. If we say Observations 1 and 2 are dependent, we mean that the probability associated with the potential events of data generating process 2 is dependent on the events from data generating process 1 (or vice versa). Dependence is then defined as in the preceding product space section.

When sampling a single observation from each of N populations, we might model each individual sampling process i by a measurable space $(\Omega_i, \mathcal{A}_i)$ in which Ω_i is the population and $\mathcal{A}_i$ is the power set of Ω_i. In other words, we do not necessarily consider the samples as coming from the same population. We can combine the whole group of measurable spaces as a single measurable space $(\Omega_1 \times \ldots \times \Omega_N, \mathcal{A}_1 \otimes \ldots \otimes \mathcal{A}_N)$. However, if we do sample from the same population N times, then $(\Omega_i, \mathcal{A}_i)$ is the same for each observation; that is, each observation is from the same space $(\Omega, \mathcal{A})$. In this case, we can denote $\Omega_1 \times \ldots \times \Omega_N$ as Ω^N, and $\mathcal{A}_1 \otimes \ldots \otimes \mathcal{A}_N$ as $\mathcal{A}^N$, yielding $(\Omega^N, \mathcal{A}^N)$ as the notation for a measurable space representing the data generating process that produced a sample of size N taken from the same population. Using probability P to model the uncertainty in the data generating process, we arrive at the probability space underlying a sample of size N as $(\Omega^N, \mathcal{A}^N, P)$. Because the outcome set is the product Ω^N, it has elements $(w_1, w_2, \ldots, w_N)$, and therefore P is a measure of sets of these elements and not individual w's.

Dependence between two observations i and j is then determined by considering the structure of the probability associated with the two-dimensional subspace and corresponding *bicoordinate marginal probability*, $(\Omega_i \times \Omega_j, \mathcal{A}_i \otimes \mathcal{A}_j, P_{i,j})$. As described in the preceding section, we ask whether events in these two coordinates are dependent. For frequentist modeling, this depends on knowing the data generating process: if you do not know the data generating process, you cannot know whether observations are dependent.

Example 4.6: Sampling with Replacement

Consider two observations sampled from the same population, in the same manner, using the same sampling probabilities. Our probability space representing the two-observation outcome and data generating process is $(\Omega^2, \mathcal{A}^2, P) = (\Omega_1 \times \Omega_2, \mathcal{A}_1 \otimes \mathcal{A}_2, P)$. Our question regards how P assigns probabilities to events in $\mathcal{A}_2$ given a realization of an event in $\mathcal{A}_1$. Let us consider $\mathcal{A}^2$ to be based on power sets, so any possible outcome can have a probability associated with it and any event is the sum of the probabilities of its constituent outcomes. Then this question can be thought of as whether the probabilities of the data generating process selecting individuals from the population change, given the event that a particular individual was selected on the other engagement of the data generating process. Because this design is one of sampling with replacement using the same sampling probabilities each time, the result obtained on one occasion does not affect the sampling probabilities underlying another observation. Observations are independent. Consider equal probability sampling (i.e., each member of the outcome set has the same probability of being selected): if after one member is selected it is placed back into the outcome set and again given the same equal probability of being selected, then the probability of obtaining any member of the outcome set for the second selection has not changed.

Example 4.7: Sampling without Replacement

Suppose, instead, that once someone is taken out of the population, the sampling probability associated with that person goes to 0 and the sampling probabilities associated with the remaining individuals are adjusted so the probability of someone other than the previously sampled person equals 1. The fact that the sampling probabilities change with each sampled individual means the observations are dependent. Consider an outcome set {Fred, Lisa, Mary} and corresponding sigma-algebra $\sigma(\{\{Fred\}, \{Lisa\}, \{Mary\}\})$; the measure describing equal probability sampling for the first selection is defined such that $P(\{Fred\}) = 1/3$, $P(\{Lisa\}) = 1/3$, $P(\{Mary\}) = 1/3$. If the first selection results in Mary as the outcome and she is not placed back into the pool of possible outcomes for the second selection, then the probability for the second selection implies $P(\{Fred\}) = 1/2$, $P(\{Lisa\}) = 1/2$, $P(\{Mary\}) = 0$. If instead the first result is Fred, then the probability for the second selection is $P(\{Fred\}) = 0$, $P(\{Lisa\}) = 1/2$, $P(\{Mary\}) = 1/2$, which is different—the distribution of the second observation depends on the result of the first and therefore the observations are dependent.

Example 4.8: Nested Sampling with Replacement

Suppose we take independent samples of physicians, and then take independent samples of those physicians' patients. We have just asserted that the physicians are independently sampled with replacement, and the patients within physicians are independently sampled with replacement; does this mean the observations are independent? Let's see. What is the outcome set? The data generating process produces a physician–patient pair: Considering two observations i and j, the outcome set is the same for each $\Omega = \text{PHYS} \times \text{PTS}$ (where PHYS is the population of physicians and PTS is the population of patients). To make this clear, suppose that our population of physicians comprises only two physicians and our population of patients comprises only four patients; moreover, let us assume each physician has two patients and the patients don't see other physicians. The sampling probabilities are as follows:

$$
\begin{array}{c}
\begin{array}{cccc}
Pt1 & Pt2 & Pt3 & Pt4
\end{array} \\
\begin{array}{c} Ph1 \\ Ph2 \end{array}
\left(
\begin{array}{cccc}
p11 & p12 & 0 & 0 \\
0 & 0 & p23 & p24
\end{array}
\right)
\end{array}
$$

where the rows are physicians and the columns are patients. The probabilities across the whole table sum to 1. However, if the data generating process has generated $(ph1, pt2)$ for one outcome in a physician cluster, then the probabilities among the other outcomes in the cluster conditional on having observed $(ph1, pt2)$ are as follows:

$$
\begin{array}{c}
\begin{array}{cccc}
Pt1 & Pt2 & Pt3 & Pt4
\end{array} \\
\begin{array}{c} Ph1 \\ Ph2 \end{array}
\left(
\begin{array}{cccc}
p11^* & p12^* & 0 & 0 \\
0 & 0 & 0 & 0
\end{array}
\right)
\end{array}
$$

where, again, the probabilities sum to 1. Given that we are sampling within a particular physician, outcomes resulting in other physicians have a probability of 0. If, instead, we observe $(ph2, pt4)$, then the probabilities of another observation within that physician cluster become

$$
\begin{array}{c}
 \quad Pt1 \quad Pt2 \quad Pt3 \quad Pt4 \\
\begin{array}{c} Ph1 \\ Ph2 \end{array}
\left(
\begin{array}{cccc}
0 & 0 & 0 & 0 \\
0 & 0 & p23^* & p24^*
\end{array}
\right)
\end{array}
$$

However, the probabilities for observations not sampled within this instance of physician p1 remain the same at

$$
\begin{array}{c}
 \quad Pt1 \quad Pt2 \quad Pt3 \quad Pt4 \\
\begin{array}{c} Ph1 \\ Ph2 \end{array}
\left(
\begin{array}{cccc}
p11 & p12 & 0 & 0 \\
0 & 0 & p23 & p24
\end{array}
\right)
\end{array}
$$

because we return the previously selected physician to the population for sampling. From this, it is evident that observations within a sampled physician are dependent but observations across sampled physicians are independent, even though both the physicians and the patients within physicians are sampled independently. Why? These probabilities are based on a data generating process, the knowledge of which we must consider in identifying dependent observations. In this case, we must know which observations are nested within the same sampled physician; knowing this allows us to identify the dependent observation. Unfortunately, inspecting the data does not tell us this—we may well have obtained the same data by a random sample with replacement of four patients from PTS and just have happened to get two patients who have physician $Ph1$ and two patients who have physician $Ph2$. In this case, patients happen to share the same physician but are not nested within physician; observations are independent, yet the data would be the same.

I cannot stress enough, when using a probability space to model a data generating process, dependence is a function of that process and not a function of the data! Mistaking this point can lead to gross errors such as the misapplication of hierarchical models that account for "dependence" to "nested" data that are not from a nested data generating process. When using a probability space to model a data generating process, as is the frequentist tradition, it is meaningless to speak of nested data; it only makes sense to speak of nested data generating processes.

Challenge 4.2

Explain why the data generating process of the preceding example does not mean that all observations underlying your data are necessarily dependent for realizations having the same physician.

Challenge 4.3

Analyze stratified sampling with replacement in terms of dependence.

The preceding examples are illuminating and common, but also straightforward. Things get more complicated when we consider observational or administrative data in which we did not actually take a sample. As stated above, whether observations are dependent, when considering a frequentist interpretation of probability, is a function of the data generating process and identifying which observations are dependent requires knowledge of that process. This can be a problem. Consider the following examples.

Example 4.9: One Day at the Clinic

Suppose you are given a data set of patient records for all the patients who showed up at an urgent care clinic on a particular day. The clinic is attended by two physicians, of whom one does the early shift and the other does the late shift. Are observations dependent? Well, what is the data generating process? First, I would submit that we don't actually know. However, perhaps we can come up with a reasonable model of one. Let's allow for randomness in the world, an ontological commitment the assessment of which is outside the scope of this book. Otherwise the set of people who show up have a probability of 1 of being selected. We might assume that nature independently selects a set of people–time pairs and then sends those people off to the clinic at their selected times. In this case, the outcome set for each observation is PEOPLE × TIME and the probabilities of an observation from this set do not depend on any result of another observation. The observations are independent, and we would not statistically cluster by physician. Again, this is a case of data resulting from independent observations that share the same characteristic (physician).

Example 4.10: Another Day at the Clinic

Suppose instead that nature selects observations in sequence: first, one person–time pair is selected to go to the clinic, then another person is selected along with a duration of time to follow the first observation. In other words, probabilities associated with times for an observation change depending on the times selected for the other observations. Again, PEOPLE × TIME is the outcome space for each observation, but now the probabilities on the time component change conditional on the results of other observations. The observations are dependent.

"Wait a minute! We were simply given the data on all the patients that showed up on a given day, and now you are telling me that whether I treat observations as dependent changes with nature's data generating process that the data cannot fully inform?" Exactly! "But which model is correct?" An excellent question, and one that does not have a clear answer if you do not know the structure of the process that generated the data. Results may support different inferences depending on which data generating process you model.

Random Variables

Up to this point, I have not mentioned random variables, nor should you assume that I have been covertly discussing them: I have not. The preceding was about the process of observation, not random variables. However, given our understanding of probability spaces, we can now understand what a random variable is, what makes it random, and from where come its probability and distributional properties.

Let $(\mathbb{R}, \mathcal{B})$ be a measurable space such that $\mathbb{R}$ represents the real line and $\mathcal{B}$ is an appropriate sigma-algebra (commonly taken to be a special sigma-algebra called a Borel sigma-algebra, but we need not concern ourselves with this detail here). If we have $(\Omega, \mathcal{A}, P)$, a probability space, and a function X from Ω to $\mathbb{R}$ such that $X^{-1}(B) \in \mathcal{A}$ for all $B \in \mathcal{B}$, then, as stated in Chapter 3, X is a function that is measurable with respect to $\mathcal{A}$ (more concisely written as "X is measurable $\mathcal{A}$," and more precisely written as "X is measurable $\mathcal{A}/\mathcal{B}$"). Note that I did not say "X is measurable P:" that would be too restrictive. So long as $(\Omega, \mathcal{A})$ is a measurable space, then X as defined above is measurable $\mathcal{A}$, regardless of there being any measure on $\mathcal{A}$ actually specified. The function X, as described here, is called a *random variable*. The triple $(\mathbb{R}, \mathcal{B}, P_X)$, with $P_X(B) = P(X^{-1}(B))$, is a probability space representing the random variable X defined on Ω having a distribution determined by P. Note in this case the subscript X on P_X denotes the random variable that the probability is modeling, not a conditional or marginal probability as has been used previously.

The key to understanding the meaning of a random variable is to understand that its distribution is directly derived from the probabilities in the model $(\Omega, \mathcal{A}, P)$—it is not necessarily the normalized frequency of a variable across the outcome space (e.g., the population being sampled), nor is it necessarily the normalized frequency of the data.

Challenge 4.4

Suppose the expected value of a random variable does not correspond to the algebraic mean of the variable in the populations: Is the sample mean an unbiased estimator of the expectation of the random variable?

We are now in a position to understand the connection between random variables and underlying probability spaces. Let's do this by example.

Example 4.11

Suppose I wish to randomly sample a person from the US population and measure his or her weight. Further, suppose I have a sampling frame for the whole population and decide to engage an equal probability data generating process. I can model this in the following way. Let Ω represent the set of individuals in the United States. I take Ω as my outcome set because my sampling frame gives me access to everyone in the United States and hence all members are potential outcomes from my data generating process. Let $\mathcal{A}$ be the power set of Ω as my sigma-algebra, thereby letting me assign probabilities to any set of outcomes. Now, to reflect the equal probability sampling data generating process, I assign $P(\{w_n\}) = 1/N$ for all singletons $\{w_n\}$ in $\mathcal{A}$ representing each of the N persons in the population. From this I get the probability of each set in $\mathcal{A}$ by additivity, because each set in $\mathcal{A}$ is just a combination of individuals and the probability of the set is the sum of the individual probabilities. For example, a set A with 2,000 people in it would have a probability of $2000/N$.

Let X be a measurement of weight on each person in the population—note that here we are speaking of a measurement and not a measure. Being a measurement, it provides an assignment of a number on the real line: therefore, $X\colon \Omega \to \mathbb{R}$. If $(\mathbb{R}, \mathcal{B})$ is a measurable space such that for all B in the sigma-algebra $\mathcal{B}$ the inverse $X^{-1}(B)$ is a set that is in the sigma-algebra $\mathcal{A}$, then X is a random variable measurable $\mathcal{A}$, and $(\mathbb{R}, \mathcal{B}, P_X)$ is a probability space with $P_X(B) = P(X^{-1}(B))$. The probability associated with the random variable is the probability that the data generating process will produce any of the members of the outcomes set that has the measurement (or set of measurements) on the random variable in which we are interested—outcomes with values for X that fall in set B.

What is the probability that the data generating process will generate a person who weighs between (and including) 100 and 110 lbs? First, we find the set of people whose weights are in that range [i.e., we identify the set A in $\mathcal{A}$ by $A = X^{-1}(\{100 \text{ to } 110\})$] and then get the probability associated with that set from the original space $(\Omega, \mathcal{A}, P)$. In other words, $P_X(\{100 \text{ to } 110\}) = P(X^{-1}(\{100 \text{ to } 110\}))$. Suppose there are n people in the subset of the population that weigh from 100 to 110 lbs; then $P_X(\{100 \text{ to } 110\}) = n/N$.

The key point here is that the probability measure of a random variable is a direct reflection of an underlying probability space. If the underlying probability space is used to model a different data generating process on the same population, we could get a different probability measure for X.

Challenge 4.5

For probability space $(\Omega, \wp(\Omega), P)$, where $\wp(\Omega)$ denotes the power set of Ω, $\Omega = \{\omega_1, \ldots, \omega_i, \ldots, \omega_{10}\}$, $P(\omega_i) = (i/55)$ and X is a random variable such that $X(\omega_i) = (11 - i)^2$.

1. What is $F(X)$ evaluated at each point in the range of X?
2. What is the probability that $20 < X < 80$?
3. What is the expected value of X?

Challenge 4.6

Show that an equal probability sample has a distribution for random variables that corresponds to the population frequency histogram of the measured variable.

Challenge 4.7

Describe a probability space for the process generating the following data, such that observations within clinics are dependent but observations across clinics are not dependent.

Clinic	Patient	HbA1c Level
1	1	8.2
1	2	7.5
2	3	6
2	4	6.6

Challenge 4.8

Describe a probability space for the process that generated the data for Challenge 4.7 such that all observations are independent.

Challenge 4.9

Describe a probability space for a process in which the observations are dependent but a random variable X is not dependent (here you will have to describe the random variable as well as the underlying probability space).

Cumulative Distribution Functions

Suppose we have a random variable X measurable $\mathcal{A}/\mathcal{B}$ that allows us to define a measure space on the real line:

$$(\Omega, \mathcal{A}, P) \xrightarrow{X} (\mathbb{R}, \mathcal{B}, P_X) \tag{4.2}$$

If X has a meaningful ordering (i.e., X is at least an ordinal-level variable) and $\mathcal{B}$ includes the sequence of intervals on the real line defined as $G = \{(-\infty, x]\}$ for all x, then the function $F_X(x) = P_X(\{(-\infty, x]\})$ is called the *cumulative distribution function* (cdf) or sometimes simply the *distribution function*. Note that this argument applies to related probability measures as well—in particular, conditional probabilities can have cumulative distribution functions. The important distinction between F_X and P_X is that the former is a function of the range of X, whereas the latter is a function of the sets in $\mathcal{B}$. As a practical matter, it is much easier to work with a function of values on the real line (e.g., F_X) than to work with a set function (e.g., P_X).

Challenge 4.10

Show that the cdf contains all the information regarding P_X and is therefore a concise means of representing the distribution of such a random variable.

Probability Density Functions

Probability measures for random variables can have corresponding density functions. Given a probability space $(\mathbb{R}, \mathcal{B}, P_X)$ for the random variable X, if there is another measure space $(\mathbb{R}, \mathcal{B}, \lambda)$ and nonnegative function f of $\mathbb{R}$, such that $P_X(B) = \int_B f \cdot d\lambda$ for all sets B in $\mathcal{B}$, then f is the probability density function for P_X with respect to λ (often called the *probability density* or just the *density* if the context makes this clear). For continuous random variables the measure λ is typically taken to be the Lebesgue measure—a detail that is unimportant for the scope of this text. Once again we have a means of arriving at probabilities in terms of functions of the range of X, in this case the density function.

If P_X has density f, then since $F_X(x) = P_X(\{(-\infty, x]\})$ and since the Lebesgue measure of an interval on the real line is simply its length that we denote as $|dx|$ for the differential dx, for continuous random variables we can write the cdf as $F_X(x) = \int_{(-\infty, x]} f(t) \cdot |dt|$. Alternatively, with a slightly different definition of integral (called the *Riemann integral*) than what we are using here, the common representation found in many textbooks would represent the integral as $F_X(x) = \int_{-\infty}^x f(t) \cdot dt$.

As with the cdf, the probability density is a function of X and not a function of sets in the corresponding sigma-algebra, which makes our modeling of distributions easier. Working with mathematical functions of values on the real line rather than working with functions of sets in sigma-algebras allows us to draw on the mathematical skills taught in the typical undergraduate setting.

Challenge 4.11

Show that for a discrete random variable the function $f(x) = P_X(\{x\})$ is a density for P_X with respect to λ defined as a count measure (i.e., $\lambda(A)$ is equal to the number of elements in the set A). In this case, the density $f(x)$ is called a *probability mass function*, and it is often not distinguished from $P_X(\{x\})$; however, the density $f(x)$ and probability measure $P_X(\{x\})$ are different. Explain how.

Expected Values

In terms of undergraduate calculus and the Riemann–Stieltjes integral, an expected value of a continuous random variable X (which is a general statement as a function of random variables is a random variable) is expressed as an integral with respect to its distribution function F:

$$E(X) = \int_{-\infty}^{\infty} x \cdot dF(x) \tag{4.3}$$

And because in this context the differential dF is the density times the infinitesimal differential dx,

$$dF = F(X + dx) - F(x) = \frac{F(x + dx) - F(x)}{dx} \cdot dx = f(x) \cdot dx \tag{4.4}$$

the expected value is also commonly expressed as the Riemann integral

$$E(X) = \int_{-\infty}^{\infty} x \cdot f(x) \cdot dx \tag{4.5}$$

In this text, it will be convenient, however, to consider expected values in terms of the underlying probability space and the measure-theoretic definition of integral provided in Chapter 3. Consider a random variable X defined on $(\Omega, \mathcal{A}, P)$. Because X is a measurable function of Ω, we can integrate it with respect to the measure P as shown in the section "Integration" in Chapter 3. If $\mathcal{A}$ is the power set of Ω, which thereby contains the set of singletons as the most granular partition Ω in $\mathcal{A}$, then in accordance with the definition of integral in Chapter 3 we would express the expected value of X as follows:

$$E(X) = \int_{w \in \Omega} X(w) \cdot P(\{w\}) \tag{4.6}$$

Expressing expected values of random variables in terms of integrals with respect to the underlying probability space, as in Equation 4.6, will become very helpful for understanding the implications of the complex data generating processes presented in Chapter 6.

Random Vectors

What do I have if I take more than one measurement on my study subjects? This is just an extension of the definition of random variables. Given a probability space $(\Omega, \mathcal{A}, P)$, a vector-valued function from an outcome space Ω to $\mathbb{R}^m$, where m is the number of components of the vector and thereby defines the number of dimensions of the real-valued space $\mathbb{R}^m$, is a random vector if there is a sigma-algebra on $\mathbb{R}^m$ (e.g., a generated sigma-algebra from the product of components $\mathcal{B}$) for which each included set can be identified with a set in $\mathcal{A}$ by an inverse set function.

Each component of a random vector is a random variable. Suppose we define a vector-valued function of Ω as follows:

$$X(w) = \begin{pmatrix} Height(w) \\ Weight(w) \\ SystolicBp(w) \end{pmatrix} \qquad (4.7)$$

For each w in Ω, X assigns three values, one each for the person's height, weight, and systolic blood pressure. Individually, *Height*, *Weight*, and *SystolicBp* are functions of Ω, measurable $\mathcal{A}$, and are each therefore random variables with ranges in the real line (perhaps some subset thereof). The vector-valued function X is therefore a random vector with a range in a space defined by three dimensions, each dimension being the real line (or some subset thereof).

The definition of a random vector is parallel to the definition of a random variable, except we can now allow for dependence across random variables in the random vector. By considering the marginal or conditional probabilities across the distribution of random variables in the random vector, we can consider the probability distributions for each component of the vector (i.e., each random variable) and relationships between the components of the random vector (e.g., correlations, regression equations, multivariate distribution models).

If we extend this logic further to define the full product space associated with a sample size of N, then for each observation we have a random vector of, say, m random variables, and we get an $N \times m$ random matrix comprising $N \cdot m$ random variables. Rows are random vectors representing possible measurements of characteristics of observations, whereas columns are random vectors representing possible measurements of a particular characteristic across observations.

This is a good place to pause and further consider what our data represent if our underlying probability space is modeling a data generating process. First, consider what the data are not: Imagine our data in an observation (rows) by variable (columns) format; the columns of our data table are *not* random variables, and the data themselves are *not* random variables. It is each individual "blank" cell of our table that is representing a random variable, with some distribution of possible values that could fill it (see Figure 4.2).

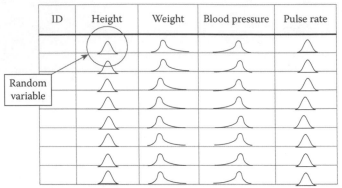

40 random variables in this matrix.

FIGURE 4.2
Random variables of a data generating process are represented by the distribution of possible values that can fill a cell in a table, not the actual data value that ends up in the cell.

Consequently, the blank table represents a random matrix, and the datum in each cell is just a particular realization of the random variable from the data generating process that underlies each cell.

This makes the concept of dependence clearer. It makes little sense to say the data are dependent: how is it that one data point, say "7," is dependent on another data point, say "33"? However, it does make sense to say that two random variables in our random matrix are dependent if the distribution of possible values that could fill one cell is different across the possible values that could fill the other cell.

Moreover, our differentiating data from the distribution of possible data provides some clarity about descriptive sample statistics and inference statistics: Descriptive sample statistics are mathematical descriptions of the data (the numbers you actually have), although data can also be modeled with a probability space (see the "Data Models" section in Chapter 5). Inference statistics are usually estimates of parameters associated with the distribution of random variables defined on probability spaces that model data generating processes.

An important consequence of understanding that each blank cell in the table represents a random variable is to note that once you run the data generating process, you only ever get one realization, one datum, for any given random variable (i.e., you only ever put one number in a cell). If you collect blood pressure measurements on a random sample of 100 people, you have one realization for each of the 100 different random variables. However, if you only have one data point for each random variable, how can you presume to have enough information to inform your understanding of each random variable's distribution? The key to accruing information is to have multiple random variables that have the same distributional characteristics.

For example, if our sample of 100 people represents realizations of 100 random variables each from an implementation of the same type of data generating process on the same population, then we have realizations from 100 random variables that each has the same distribution for blood pressure. In this case, each datum informs the same distribution, even though each is a realization from a different random variable. It is important to note the distinction between having only one realization per random variable and having multiple random variables (and thereby multiple realizations) that have the same distributional characteristic of interest.

This notion, as I've stated it, may at first seem foreign, but it is implicit in the classic manner in which the sample mean is shown in introductory statistics classes to be an unbiased estimator of a random variable's expected value. Suppose we have a data set comprising measurements on some characteristic X for n individuals (i.e., the sample size is n). The sample mean statistic, $\overline{X}$, is defined as follows:

$$\overline{X} = \frac{1}{n} \cdot \sum_{i=1}^{n} X_i \qquad (4.8)$$

The expected value of the sample mean is

$$E(\overline{X}) = E\left(\frac{1}{n} \cdot \sum_{i=1}^{n} X_i\right) = \frac{1}{n} \cdot \sum_{i=1}^{n} E(X_i) \qquad (4.9)$$

The very fact that the expected value of the sample mean is expressed as a function of the sum of the expected values for each X_i (see the rightmost term of Equation 4.9) implies that each of the n X_i's is taken to be a distinct random variable with its own expected value. Moreover, if each of the n random variables has the same expected value, say μ, then

$$E(\overline{X}) = \frac{1}{n} \cdot \sum_{i=1}^{n} E(X_i) = \frac{1}{n} \cdot \sum_{i=1}^{n} \mu = \frac{1}{n} \cdot (n \cdot \mu) = \mu \qquad (4.10)$$

Consequently, the expected value of the sample mean, as a function of n random variables having the same expected value, is the expected value of those individual random variables. Therefore, the sample mean is considered an unbiased estimator of the expected value of X_i for all i. This conclusion, regarding the unbiased nature of the sample mean, assumes the very ideas presented above; specifically, each X_i is a distinct random variable, and these random variables share a common distributional characteristic—in this case, they each have the same expected value.

To press this point further, if we have an $n \times m$ random matrix (e.g., the potential outcomes of m measures on n observations), comprising by necessity $n \cdot m$ random variables, without further knowledge we cannot combine the resulting data to inform questions regarding underlying distributions.

Why? Because without further knowledge, we simply have a single realization from each of $n \cdot m$ completely different distributions, that is, distributions not presumed to share any characteristic of interest. We need to presume, by virtue of design or reasonable assumption, that some random variables share common characteristics in their distributions to warrant combining their corresponding data.

Dependence within Observations

Suppose we have a random vector with m elements defined on $(\Omega, \mathcal{A}, P)$, then we simply have m random variables, which are measurements on the elements of our outcome set Ω.

Example 4.12

Let Ω denote the set of individuals in a population, then the measurements of height and weight associated with each individual compose a two-component random vector of (Height(w), Weight(w)) that comprises the random variables Height(w) and Weight(w), assuming of course that they are measurable with respect to $\mathcal{A}$.

To understand whether such random variables are independent or dependent requires us to tie them back to the independence and dependence of events in $\mathcal{A}$, the sigma-algebra associated with the underlying probability space. This should not be surprising because the probabilities associated with random variables derive from the underlying probability space; as such, everything we do with our random variables must tie back to the original probability space. We take this step by noting that each random variable X generates a partition π_X on the outcome space, which can be used to generate a sigma-algebra $\sigma(\pi_X)$ that is contained in $\mathcal{A}$, assuming $\mathcal{A}$ is rich enough to support the partition. The sigma-algebra $\sigma(\pi_X)$ is termed a *sub-sigma-algebra* of $\mathcal{A}$.

Example 4.13

Let $(\Omega, \mathcal{A}, P)$ represent a data generating process that samples from a population Ω of people. Let an indicator of being female be a random variable Female associated with a given population. Female assigns two possible values: 1 if the individual is female and 0 if the individual is male. The sets Female$^{-1}(\{1\})$ and Female$^{-1}(\{0\})$ in $\mathcal{A}$ are subsets of Ω that compose a partition of Ω (everyone in Ω is a member of either in Female$^{-1}(\{1\})$ or Female$^{-1}(\{0\})$ but not in both). The sub-sigma-algebra of $\mathcal{A}$ generated by the Female partition is $\sigma(\text{Female}^{-1}(\{1\})) = \{\Omega, \varnothing, \{\text{female}\}, \{\text{male}\}\}$.

Challenge 4.12

Explain why the sigma-algebra generated from the full partition is the same as the sigma-algebra generated from all sets in the partition but one (any one).

Random variables defined on $(\Omega, \mathcal{A}, P)$ are independent if their corresponding sub-sigma-algebras are independent. In other words, the events associated with one sigma-algebra are each independent of the events in the other; the independence of events was described above. What this means is that random variables X and Y are independent if for all events $A \in \sigma(\pi_X)$ with positive measure and all events $B \in \sigma(\pi_Y)$, $P(B \mid A) = P(B)$, or generally defined as $P(A \cap B) = P(A) \cdot P(B)$. Another way to think of this is that the trace of $\sigma(\pi_Y)$ on A should produce the same relative "sizes" (in a Venn diagram sense) across the events in $\sigma(\pi_Y)$ for all $A \in \sigma(\pi_X)$. No matter which event A that I consider, the relative probabilities of events B remain the same.

Example 4.14

Suppose we wish to know whether a variable that indicates men versus women is independent of a variable that indicates being right-handed versus left-handed (assuming no one is ambidextrous). The top row of Figure 4.3 indicates the case in which the partition on the population generated by the sex variable and the partition generated by the handedness variable produce independent events. Note that the proportion of the circle that is right-handed is 0.5, and the proportion of right-handed

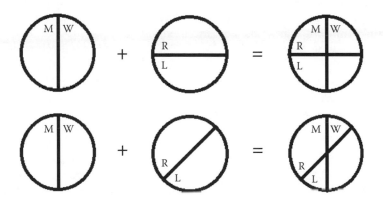

FIGURE 4.3
Independent and dependent variables: the top row depicts independent variables; the proportion of the area of the circle captured by values of one variable remains the same regardless of the value of the second variable (e.g., R takes up half of the area of both M and W). The bottom row depicts dependent variables; the proportion of the area captured by values of one variable depends on the value of the other variable (e.g., R takes up a larger portion of M than it takes up of W).

given women is also 0.5, as is the proportion of right-handed given men. The bottom row of Figure 4.3, however, shows a case in which the events are not independent. The same overall proportions hold for Sex and Handedness; however, now the proportion of right-handed given women is less than 0.5, whereas the proportion of right-handed given men is greater than 0.5. The conditional probabilities do not equal the marginal probabilities—Sex and Handedness are dependent.

Note that, assuming $\mathcal{A}$ is rich enough to support as random variables all the measurements we want, the fact that some pairs of random variables are independent and others are not independent shows that just because events in $\mathcal{A}$ are not generally independent does not mean that certain sub-sigma-algebras (ways of chopping up Ω) are not independent of other sub-sigma-algebras (other ways of chopping up Ω).

Dependence across Observations

Above I established that for frequentists the data generating process dictates whether observations are dependent. Does our determination of observational dependence translate over to random variables defined across observations? First I'll let the cat out of the bag and then I'll explain: Random variables for independent observations are necessarily independent across observations. However, random variables for dependent observations *may or may not be dependent* across observations. The consequence of this latter point is that knowing observations are independent immediately tells us that the corresponding random variables are independent across those observations, whereas knowing that observations are dependent does not tell us whether our random variables are dependent, only that they may be. Therefore, just because we use a clustered sampling design, which, as we have seen, gives us dependent observations, does not mean that the random variables of interest are dependent across observations.

Example 4.15

Suppose a data generating process was a cluster sampling design in which physicians are sampled and then patients are sampled within physician. The (physician, patient) observations are dependent within physician as discussed above. However, suppose that the distribution of patient ages is the same for each physician in the physician population being sampled. This being the case, the conditional probability for patient age (a random variable) does not change given the result of the patient age on another dependent patient observation. By contrast, suppose the distribution of HbA1c differs across physicians; consequently, across dependent observations, the resulting HbA1c for one observation informs the distribution of HbA1c for the other, and the corresponding random variables are thereby dependent.

Example 4.16

Suppose we have two physicians labeled Ph1 and Ph2 and six patients labeled pt1, pt2, pt3, pt4, pt5, and pt6; further suppose that patients pt1, pt2, and pt3 see only physician Ph1, whereas patients pt4, pt5, and pt6 see only physician Ph2. Our set of outcomes for a single sampling instance can be specified as $\Omega = \{(\text{Ph1}, \text{pt1}), (\text{Ph1}, \text{pt2}), (\text{Ph1}, \text{pt3}), (\text{Ph2}, \text{pt4}), (\text{Ph2}, \text{pt5}), (\text{Ph2}, \text{pt6})\}$. Let $\mathcal{A}$ be the power set of Ω, and P be a probability measure representing the sampling probabilities for a clustered sampling strategy (physicians are sampled first, then patients within physicians—each with replacement). Our measure space for a sample size of N is $(\Omega^N, \mathcal{A}^N, P)$. Note that we could just as well specify our outcome set as all physician–patient pairs and designate P to assign a zero probability to any pair that doesn't match the above combinations, but there is no advantage to this here.

From previous examples we know the observations within physicians are dependent, so let's consider an arbitrary bicoordinate subspace representing two dependent observations, say Observations 1 and 2. Our measure space of interest here is then $(\Omega_1 \times \Omega_2, \mathcal{A}_1 \otimes \mathcal{A}_2, p)$. Our concern is whether random variables measured on this space are independent or dependent.

Note that an arbitrary element w of the outcome space contains two physician–patient pairs, one for Observation 1 and the other for Observation 2: $w = ((\text{Ph}_i, \text{pt}_j)_1, (\text{Ph}_k, \text{ph}_l)_2)$ with i and $k \in \{1, 2\}$, j and $l \in \{1, 2, 3, 4, 5, 6\}$. Let's define two functions on the outcomes set $\Omega_1 \times \Omega_2$. One indicates whether the patient on Observation 1 weighs less than 150 lbs; the other indicates whether the patient on Observation 2 weighs less than 150 lbs:

$$X_1(w) = \mathbf{1}_{wt((\text{pt}_j)_1) \ < \ 150lbs}(w) \tag{4.11}$$

$$X_2(w) = \mathbf{1}_{wt((\text{pt}_j)_2) \ < \ 150lbs}(w) \tag{4.12}$$

Suppose the functions in Equations 4.11 and 4.12 map to the measure space $(\mathbb{R}^2, \mathcal{B}^2)$, with $\mathcal{B}^2 = \mathcal{B}_1 \otimes \mathcal{B}_2$, such that they are measurable $\mathcal{A}_1 \otimes \mathcal{A}_2$. They are, therefore, random variables and we can consider their probabilities p_X as determined by p. $(\mathbb{R}^2, \mathcal{B}^2, p_X)$ is a probability space representing the distribution of the random variables associated with the underlying data generating process.

If X_1 and X_2 are independent, then $p_X(B_2 \mid B_1) = p_X(B_2)$ for all $B_1 \in \mathcal{B}_1$ with positive measure and $B_2 \in \mathcal{B}_2$. As noted in the preceding section, the independence of random variables derives from the independence of their associated sub-sigma-algebras of events in the underlying probability space. What is the event associated with $X_1 = 1$? Suppose each patient has weight less than 150 lbs as indicated in Table 4.3, in which 1 = less than 150 lbs and 0 = greater than or equal to 150 lbs.

TABLE 4.3

Values for Indicator X of Weight Less Than 150 lbs

	Ph1			Ph2		
	Pt1	**Pt2**	**Pt3**	**Pt4**	**Pt5**	**Pt6**
$1_{wt<150}$	1	0	0	1	1	0

TABLE 4.4

Pairs of Observations for which $X_1 = 1$ in the Case of Different Distributions across Physicians

			Observation 1					
			Phys 1			Phys 2		
			1	**2**	**3**	**4**	**5**	**6**
Observation 2 — Phys 1		1	1					
		2	0			✕		
		3	0					
Phys 2		4				1	1	
		5	✕			1	1	
		6				0	0	

Patients 1, 4, and 5 have low weight, whereas Patients 2, 3, and 6 have high weight. Notice that the distributions of weight in the patient populations across physician are *not* the same. In this case, what happens if we condition on the case of $X_1 = 1$, that is, Observation 1 produces a patient with low weight. The cells containing numbers in Table 4.4 indicate all of the outcomes in $\Omega_1 \times \Omega_2$ for which this condition is true. There are nine possible outcomes for which X_1 is 1—remember that each outcome contains two physician–patient pairs and we are conditioning on the first pair.

There are five out of nine possible outcomes having a patient in the second observation with low weight, that is, five possible outcomes with $X_2 = 1$. Now let's consider when we condition on $X_1 = 0$. The columns containing numbers in Table 4.5 represent the outcomes meeting this condition. Again, there are nine possible outcomes that meet the condition, but only four out of nine have the patient in the second observation with low weight. For the two preceding tables, we see that the probability of X_2 depends on X_1; therefore, X_1 and X_2 are dependent.

Instead, suppose each patient has weight less than 150 lbs as indicated in Table 4.6. Patients 1 and 4 have low weight, whereas Patients 2, 3, 5, and 6 have high weight. Note that the distributions of weight in the patient populations across physician are the same. In this case, what happens if we condition on the case of $X_1 = 1$, that is, Observation 1 produces a patient with low weight? The cells with numbers in Table 4.7

TABLE 4.5

Pairs of Observations for which $X_1 = 0$ in the Case of Different Distributions across Physicians

			Observation 1					
			Phys 1			Phys 2		
			1	2	3	4	5	6
Observation 2	Phys 1	1		1	1			
		2		0	0			
		3		0	0			
	Phys 2	4						1
		5						1
		6						0

TABLE 4.6

Values for Indicator X of Weight Less Than 150 lbs

	Ph1			Ph2		
	Pt1	Pt2	Pt3	Pt4	Pt5	Pt6
$\mathbf{1}_{wt<150}$	1	0	0	1	0	0

TABLE 4.7

Pairs of Observations for which $X_1 = 1$ in the Case of Same Distributions across Physicians

			Observation 1					
			Phys 1			Phys 2		
			1	2	3	4	5	6
Observation 2	Phys 1	1	1					
		2	0					
		3	0					
	Phys 2	4				1		
		5				0		
		6				0		

indicate all of the outcomes in $\Omega_1 \times \Omega_2$ for which this condition is true. There are six possible outcomes for which X_1 is 1—remember that each outcome contains two physician–patient pairs.

One-third of the possible outcomes have the patient in the second observation with low weight, that is, with $X_2 = 1$. Now consider conditioning on $X_1 = 0$. Table 4.8 represents the twelve outcomes meeting this

TABLE 4.8

Pairs of Observations for which $X_1 = 0$ in the Case
of Same Distributions across Physicians

			Observation 1					
			Phys 1			Phys 2		
			1	2	3	4	5	6
	Phys 1	1		1	1			
		2		0	0			
Observation 2		3		0	0			
		4					1	1
	Phys 2	5					0	0
		6					0	0

condition. Again, one-third of the possible outcomes have the patient in the second observation with low weight. Under both conditions of X_1, the probability of X_2 remains the same: X_1 and X_2 are independent.

In both scenarios in Example 4.16, the underlying probability space was the same, representing the dependent observations of a cluster sampling design; however, the random variables in the first scenario were dependent, whereas the random variables in the second scenario were independent. For this kind of study design, independence occurs within cluster when the distribution of the variables is the same in each cluster.

Another View of Dependence

Remember the definition of a partition and note that sigma-algebras contain partitions. The simplest example of this claim is that for any set A in a sigma-algebra $\mathcal{A}$ on outcome set S, $\mathcal{A}$ also includes $\overline{A}$, the complement of that set by definition; the pair A and $\overline{A}$ is a partition of the outcome set S.

Moreover, if a partition Π is contained in the sigma-algebra $\mathcal{A}$, then the sigma-algebra generated by the partition $\sigma(\Pi)$ is contained in $\mathcal{A}$. Again, this should be clear because $\mathcal{A}$ must include all the complements and countable unions of its sets and therefore for any collection of sets in $\mathcal{A}$, $\mathcal{A}$ must include all the sets in the corresponding sigma-algebra generated from that collection. Because a partition of Ω in $\mathcal{A}$ is a collection of sets, the sets in the sigma-algebra generated by the partition are also in $\mathcal{A}$.

For a probability space $(\Omega, \mathcal{A}, P)$ with a partition Π of Ω contained in $\mathcal{A}$, the space $(\Omega, \sigma(\Pi), p)$ is a probability space with $p = P$ for all $A \in \sigma(\Pi)$. In other words, we don't change the numerical assignment of probability; we merely cull from $\mathcal{A}$ a sub-sigma-algebra generated by a particular partition. In mathematical terms, we say that p is the *restriction* of P to the measurable space $(\Omega, \sigma(\Pi))$. It should not be surprising that we can do this, because a partition of Ω is a collection of disjoint sets that cover Ω and the sum of P across the sets that make up Ω is 1.

Suppose we wish to consider the dependence status of two random variables associated with observations i and j from $(\Omega^N, \mathcal{A}^N, P)$, which is to say we consider random variables defined on $(\Omega_i \times \Omega_j, \mathcal{A}_i \otimes \mathcal{A}_j, p)$. The dependence status of random variables across observations depends on whether the observations are dependent. Remember that observations are dependent if

$$\exists A_i \in \mathcal{A}_i \, \exists \, A_j \in \mathcal{A}_j (p(A_j|A_i) \neq p(A_j)) \tag{4.13}$$

This is to say that the distribution of events on observation j is not the same as the distribution conditional on some event A_i. For simplicity of notation, take A_i and A_j to be a pair of events for which the preceding inequality holds. Then, it must be the case that $p(A_j|A_i) \neq p(A_j|\overline{A}_i)$; therefore, the sets of the partition $(A_i, \overline{A}_i)$ are contained in $\mathcal{A}_i$ across which the conditional distributions of events in $\mathcal{A}_j$ differ.

More generally, suppose we have a partition $\Pi = (\pi_1, \pi_2, \ldots)$ contained in $\mathcal{A}_i$ such that the following two conditions hold:

1. $p(A_j|\pi_k) \neq p(A_j|\pi_m)$ for all distinct pairs π_k and π_m in Π.
2. $p(A_j|B) = p(A_j|\pi)$ for all $B \in \mathcal{A}_{i\pi}$ and for all $\pi \in \Pi$, where $\mathcal{A}_{i\pi}$ is the trace of $\mathcal{A}_i$ on π.

Although Condition 1 is overly strict, as we only need two sets of the partition for which the probabilities are not equal, this specification is useful in identifying the minimal partition that drives dependence. Condition 1 states that observations i and j are dependent by virtue of events in the partition Π contained in $\mathcal{A}_i$; if there were sets in the partition for which Condition 1 did not hold, we would simply replace them with their union to get another partition for which the condition did hold. Condition 2 states that events are independent within each set of the partition. If Condition 2 does not hold for any set π of the partition Π, then we subdivide that π further to achieve a finer-grained partition until Condition 2 is met (assuming the algebra is sufficiently rich). One way to think of this is that the nature of dependence across observations is characterized completely by the partition Π of Ω_i that is contained in $\mathcal{A}_i$.

Let Z be a random variable that assigns to each element of the outcome set a number identifying the set of the partition Π to which it belongs. Moreover, assume there are at least two sets in the partition with nonzero probabilities. Because Π characterizes the dependence structure, we should expect that Z, which numerically indexes Π, does so as well. This is indeed the case.

Let X_i and X_j be random variables: are they dependent? The joint probability of these two random variables is

$$p(X_i, X_j) = \sum_{Z_i} p(X_i, X_j | z_i) \cdot p(z_i) \tag{4.14}$$

which by Condition 2 above, which states that X_i and X_j are independent conditional on each z_i, is

$$p(X_i, X_j) = \sum_{Z_i} p(X_i | z_i) \cdot p(X_j | z_i) \cdot p(z_i) \tag{4.15}$$

By definition, if X_i and X_j are independent, then Equation 4.15 must reduce to

$$p(X_i, X_j) = p(X_i) \cdot p(X_j) \tag{4.16}$$

otherwise the random variables are dependent.

If the distribution of either one of the random variables does not change across Z_i, then X_i and X_j are independent. For example, suppose the distribution of X_j does not vary across the partition characterizing the dependence between the observations [i.e., $p(X_j | z_i) = p(X_j)$], but the distribution of X_i does vary [i.e., $p(X_i | z_i) \neq p(X_i)$]. Then

$$p(X_i, X_j) = p(X_j) \cdot \sum_{Z_i} p(X_i | z_i) \cdot p(z_i) = p(X_j) \cdot p(X_i) \tag{4.17}$$

Moreover, X_i and X_j are independent. If neither of the random variables X_i and X_j is independent of z_i, then the two random variables are dependent and the equation

$$p(X_i, X_j) = \sum_{Z_i} p(X_i | z_i) \cdot p(X_j | z_i) \cdot p(z_i) \tag{4.18}$$

gives us insight into the reason for dependence: random variables are dependent when their distributions vary across the partition indexed by the random variable Z that characterizes the dependence between observations.

Note, however, that for a random variable to be dependent on the indicators of a partition, it is not required that the distribution of the random variable be different across all components of the partition. It is only required that there exists at least one component of the partition for which the distribution of the random variable is different. This will be important when analyzing the dependent structure of study designs such as the cluster-randomized clinical trial in which clusters are sampled, the whole cluster is assigned to a treatment or control condition, and individuals within the cluster are sampled for measurement (or all individuals sampled within a cluster are assigned a treatment as designated by the assigned cluster-level treatment designation).

A few paragraphs above I mentioned in the condition, "Moreover, assume there are at least two sets in the partition with nonzero probabilities." What if instead there is a particular z for which $p(z) = 1$? Well, then the preceding equation becomes

$$p(X_i, X_j) = p(X_i|z) \cdot p(X_j|z) \tag{4.19}$$

Equation 4.19 doesn't look exactly like our criterion for independence: are X_i and X_j independent? They are if $p(X_i|z) = p(X_i)$ and $p(X_j|z) = p(X_j)$, which is indeed the case. This is easy to show, so I will leave it as a challenge.

Challenge 4.13

Show for the preceding case that indeed $p(X_i|z) = p(X_i)$ and $p(X_j|z) = p(X_j)$ if there is a particular z with $p(z) = 1$. Hint: this is easily shown intuitively by Venn diagrams on the underlying probability space.

Example 4.17

For a cluster sample in which patients are nested within physicians, our partition is determined by physician, and z_i is a variable that identifies the physician for observation i. The data generating process, by virtue of being a cluster sampling process, is such that z_i also indicates the physician of observation j. Consequently, if the distribution of the qualities measured by the X's varies across physician, then the random variables are dependent. Note that the random variables X_i and X_j need not measure the same qualities on the two observations. Preceding examples used weights on each patient; however, it could just as well be observation i's weight and observation j's patient satisfaction. If both vary by physician, then they are dependent.

The key point of this section is that random variables are not dependent if the observations are not dependent, and random variables *may* be dependent if the observations are dependent. If there is a partition on the outcome set across which the conditional distributions of random variables differ, then the random variables are dependent. A corollary to this last point is that the intraclass (or intracluster) correlation does not determine whether observations are dependent, only whether random variables are dependent given the *a priori* determination that the observations are dependent. If we know that the data generating process is such that the observations are independent, the corresponding random variables are independent regardless of the intraclass correlation, and the intraclass correlation should not be taken as evidence for dependence.

Densities Conditioned on Continuous Variables

It is common to speak of conditional probabilities, densities, and moments as if they were conditioned on a single point in a continuous space. For example, treating weight as a continuous variable, we might say the probability density of systolic blood pressure given weight equal to 210 pounds has a particular distribution, perhaps a normal distribution. However, our measure-theoretic understanding of random variables suggests that such statements are wrong. The easiest way to see this is by considering the following two equations regarding the joint distribution of two random variables Y and X, where B denotes a measurable set in the range of Y and x is a single point:

$$P(B|\{x\}) = \int_{y \in B} f(y|x)dy \qquad (4.20)$$

and

$$P(B|\{x\}) = \frac{P(B, \{x\})}{P(\{x\})} \qquad (4.21)$$

Equation 4.20 simply establishes f as what we would like to consider the usual conditional probability density of Y, given that X is equal to some value x that can produce the probability on the left-hand side of the equation. However, Equation 4.21 shows that this conditional probability is in fact not defined because the probability measure of a point in a continuous space is usually considered to either be zero or not defined (depending on the measure of the corresponding density). Consequently, the probability in the denominator of the right-hand side of the second equation, $P(\{x\})$, is zero or not defined and consequently neither is the conditional probability on the left-hand side. Therefore, even though the function f on the right-hand side of the first equation is nonnegative, integrates to 1, and calculates to a number, it cannot be the desired conditional probability density, because, as stated, the conditional probability it purports to support does not actually exist. Therefore, the first equation above is not actually a meaningful equation.

To add to the confusion, which I will clear up below, the joint probability density $f(y, x)$ and marginal densities $f(y)$ and $f(x)$ are well defined, as is the equation $f(y, x) = f(y|x) \cdot f(x)$. The conditional function is the ratio of two probability densities, $f(y|x) = \frac{f(y, x)}{f(x)}$, and is itself a density for some measure on Y, X, or both, just not the probability measure we might presume for the conditional probability of interest. However, clearly our disputed function $f(y|x)$ serves a purpose. So, what is it? Well, though $f(y|x)$ is not the presumed conditional probability density function, it is a function of Y and X that relates a marginal density to a joint density.

This is all well and good when we are using $f(y|x)$ to connect a marginal distribution to a joint distribution, but we use such conditional "probability densities" in applied research for more than specifying joint distributions (specifically, researchers are often interested in the parameters of conditional distributions). For example, they are presumed to underlie regression functions such as

$$E(Y|X=x) = \int y \cdot f(y|x)dy \qquad (4.22)$$

However, the fact that $f(y|x)$ is not a probability density should lead us to suspect that the function $E(Y|X=x)$ is not actually a conditional expectation—indeed, as argued above, it is not! What we should be speaking of is $E(Y|X \in \Delta_x)$, where Δ_x denotes a measurable set of points containing x with positive measure.

To understand the role of $f(y|x)$ in quantities of interest, consider the conditional probability $P(A|\Delta_x)$, where A is a measurable set associated with Y, and again Δ_x denotes a measurable set of points containing x for which $P(\Delta_x)$ is not zero:

$$P(A|\Delta_x) = \frac{P(A, \Delta_x)}{P(\Delta_x)} \qquad (4.23)$$

This is a well-defined equation with the left-hand side a proper conditional probability statement. The numerator on the right-hand side can be expressed as

$$P(A, \Delta_x) = \int_A \int_{\Delta_x} f(y, x)dxdy \qquad (4.24)$$

where $f(y, x)$ is the joint probability density that is the product of our disputed $f(y|x)$ and the marginal density $f(x)$:

$$P(A, \Delta_x) = \int_A \int_{\Delta_x} f(y|x) \cdot f(x)dxdy \qquad (4.25)$$

By bringing the constant $P(\Delta_x)$ under the integral, we can thereby express $P(A|\Delta_x)$ as

$$P(A|\Delta_x) = \int_A \int_{\Delta_x} f(y|x) \cdot \underbrace{\underbrace{\left(\frac{f(x)}{P(\Delta_x)}\right)}_{\text{Part 1}} dxdy}_{\text{Part 2}} \qquad (4.26)$$

where Part 1 of the right-hand side is a legitimate conditional density of X given $X \in \Delta_x$, and Part 2 is therefore the expected value of $f(y|x)$, with respect to the distribution of X given $X \in \Delta_x$ for any value of Y. Consequently,

$$P(A|\Delta_x) = \int_A E_X(f(y|x)|X \in \Delta_x)dy \qquad (4.27)$$

which, by our definition for probability densities, implies that $E_X(f(y|x)|$ $X \in \Delta_x)$ is the conditional probability density that corresponds to the conditional probability measure $P(A|\Delta_x)$. So, the conditional probability $P(A|\Delta_x)$ has a legitimate probability density, which we can label $f(y|\Delta_x)$, that is the conditional expectation of $f(y|x)$ taken with respect to the conditional distribution of X given $X \in \Delta_x$.

Now, if we consider $f(y|\Delta_x)$ with respect to a nested sequence of convex sets centered on the singleton $\{x\}$ starting from Δ_x equal to some set, say $\Delta_x{}^*$, such that the sequence converges to the singleton $\{x\}$, then $f(y|\Delta_x)$ with respect to this sequence will approach $f(y|x)$; technically, we would take the limit of this sequence, assuming the limit exists. Consequently, we can use $f(y|x)$ to approximate $f(y|\Delta_x)$ if Δ_x is a really small set with positive measure containing x, and our common use of $f(y|x)$ as if it were the probability density of interest, though technically wrong, is vindicated as an approximation in this case. Note that in empirical work Δ_x is naturally taken to be a set of X values within the unit of measurement around the measured value x—which may not be small enough for a good approximation.

Note the technical consequence of this section's discussion regarding our definitions of regression functions. Specifically, regression functions that are defined as functions of continuous random variables are in fact not actually conditional expectations; instead, they are limits of conditional expectations with respect to sequences of sets that converge to the singular set containing the point being conditioned on, which is not measurable. Nonetheless, it is often convenient to write expressions such as $E(Y|X \in \Delta_x)$ as $E(Y|X)$: I will use this convention where it is not confusing.

Statistics

Statistics is a term used to label a field of study and an activity, but formally a statistic is a type of measurable function. For example, suppose we have a data generating process underlying N observations that we model as $(\Omega^N,$ $\mathcal{A}^N, P)$. The outcome set in this example is the product of N sets: $\Omega^N = \Omega_1 \times \Omega_2 \times \dots \times \Omega_i \times \dots \times \Omega_N$, in which $\Omega_n = \Omega$ for all $n \in \{1, 2, \dots, N\}$. An individual element of Ω^N is a group of N elements, one from each component set of Ω^N: that is, $(w_1, w_2, \dots, w_N) \in \Omega^N$. Suppose we measure a characteristic X on each of the N individuals that comprise an outcome $(w_1, w_2, \dots, w_N)$; X then underlies N random variables:

$$(\Omega^N, \mathcal{A}^N, P) \xrightarrow{X_1, \dots, X_N} (\mathbb{R}^N, \mathcal{B}^N, P_X)$$

We therefore have N random variables X_1 to X_N.

Because $(\mathbb{R}^N, \mathcal{B}^N, P_X)$ is a probability space, we can define a random variable with respect to it by a measurable function from $\mathbb{R}^N$ to $\mathbb{R}$, the real line. The average of the N random variables is a common example:

$$\hat{\mu} = \frac{1}{N} \sum_{i=1}^{N} X_i \tag{4.28}$$

In this case, $(\mathbb{R}^N, \mathcal{B}^N, P_X) \xrightarrow{\hat{\mu}} (\mathbb{R}, \mathcal{B}, P_\mu)$ and $\hat{\mu}$, by virtue of P_μ, has a probability density f_μ and distribution F_μ. The function, $\hat{\mu}$, of the random variables X_1 to X_N is called a *statistic*; its probability P_μ, and consequently its distribution, reflects P from the original probability space via its connection through P_X. If the original probability space is solely modeling a data generating process, then the distribution of $\hat{\mu}$ is called its *sampling distribution*. This distribution reflects the values of $\hat{\mu}$ that are possible due to the different sets of N observations obtainable by the data generating process. The standard deviation of the distribution for $\hat{\mu}$ is its standard error. See the "Interpreting Standard Errors" section of Chapter 6 for a discussion of the relationship between standard deviations and standard errors in probability spaces that do not solely model data generating processes.

It is interesting to note that, although statistics are often defined as the sum of functions of random variables, such as the sample mean of the preceding example, it is not necessary that each individual component of the sum be a random variable. This is evident in the maximum likelihood estimator. Suppose we define two random variables, Y and X, for each of N observations. This results in $2 \cdot N$ random variables (i.e., an (X, Y) pair for each observation). For $ll(Y_i, X_i; \theta)$ denoting the log-likelihood component associated with observation i, the maximum likelihood estimator is defined as follows:

$$\hat{\theta} = \underset{\Theta}{\text{argmax}} \sum_{i=1}^{N} ll(Y_i, X_i; \theta) \tag{4.29}$$

Assuming appropriate conditions are met that guarantee a maximum for all possible outcomes and θ values, then $\hat{\theta}$ is a random variable: $(\mathbb{R}^{2N}, \mathcal{B}^{2N}, P_{YX}) \xrightarrow{\hat{\theta}} (\mathbb{R}, \mathcal{B}, P_\theta)$. However, in this case, each log-likelihood component of the sum shown above is not a random variable that maps to the real line because θ is unspecified (although $ll(Y_i, X_i; \theta)$ is a random function of θ). Nonetheless, $\hat{\theta}$ is a random variable with properties that make it useful for using the data-specific values of $\hat{\theta}$ (i.e., an estimate) to inform hypothesis tests and estimation goals (e.g., under certain conditions being consistent for the θ value that parameterizes the distribution underlying the log-likelihood).

Some reflection should reveal the distinction between the role of the likelihood function in determining the maximum likelihood statistic and the role of the likelihood in classic Bayesian analysis. As a statistic, its role is purely instrumental: as it turns out, by proof, this function has properties we desire of an estimator that make it useful, but there are often other estimators that we could just as well use. Indeed, once I know the estimator has the properties I want (e.g., being consistent and asymptotically efficient), I do not really care about the inherent meaning of the underlying function, except as required for proper specification. In a classic Bayesian analysis, however, the likelihood function is not a statistic, but rather is part of the specification of a joint probability distribution that dictates the meaning of the posterior distribution being sought. In the Bayesian analysis the likelihood provides meaning to the posterior.

What's Wrong with the Power Set?

In the preceding sections, I usually jumped straight to specifying my sigma-algebra as the power set of a finite outcome set of interest. Is this always wise? For most of our purposes as applied researchers, I argue it is useful, but first let's see why it can be a problem. Some contend that the power set can be too large to assign probabilities and therefore should not be specified when the outcome set is anything but small. This contention presupposes that you are assigning the probabilities to the events of this sigma-algebra; if true, I agree. Typically when you endeavor to make assignments to the base-level probability space (as opposed to a random variable generated space), you are doing certain types of experiments or discerning a measure of subjective uncertainty.

Example 4.18: Particle Beam Deflection

Suppose two young physicists, Matthew and Devin, hypothesize that under certain conditions a beam of particles shot toward a wall will be deflected to the left of the vertical centerline. Also suppose the wall shows where the beam hits. The outcome space for this experiment can be specified by the cells of a fine grid on the surface of the wall (say each cell is 0.000000001 inches square).

Matthew decides to take the event set to be the power set of the outcome set, and he therefore has a super-large set of events. If Matthew wishes to directly assign frequentist probabilities to the event space as operationalized via multiple independent runs of the experiment, he would need to rerun the experiment an extremely large number of times to get a good estimate of the probabilities associated with each event. Once done, he can test his hypothesis.

Devin, by contrast, realizes that her hypothesis is simply left of center versus right of center, so she specifies her event space to be $\sigma(\{\text{left of center}\}) = \{\{\text{The Whole wall}\}, \varnothing, \{\text{left of center}\}, \{\text{right of center}\}\}$. Devin needs only to track the proportion of points that are left of center to fill out her probability measure.

Devin will be home for dinner tonight; Matthew won't even make his retirement party.

When using experiments to directly assign the probability measure of an event space, the power set can indeed be too large for practical use, and careful consideration of what you are really interested in becomes a very important practical matter.

If instead we are interested in features of random variables, rather than directly identifying the probabilities of the underlying probability space, then we can simply use the data. The underlying probability space generates the random variables, the distribution of which is reflected in samples from the data generating process. So long as we use the same data generating process we suppose we are modeling, then information in the data (realizations of random variables) is sufficient to estimate and test other quantities of interest.

Example 4.19

Continuing the preceding example, Matthew may keep his specification of the event space but then measure a random variable that indicates left of center and use an estimator of the proportion of points with the value of the random variable associated with left of center. Now Matthew can go home for dinner as well.

Do We Need to Know P to Get P_X?

It may seem from the preceding sections of this book that we must actually calculate the probabilities P_X of random variables X from the underlying probability P, thereby requiring us to know P in order to determine the distribution of X. This is not so. Although the distribution of X comes from P, it is still the distribution of X and we can discern it from a sample (using data generating processes that produce a series of random variables having the same distribution) without knowing or assigning P. Or we can hypothesize a distribution for X and use data from random variables having the same distribution to test the hypothesis. In other words, nothing said so far precludes our usual approaches to modeling random variables from the data comprising realizations of random variables from a data generating process. However, our model specification and interpretation of results strongly depend on the preceding discussion.

It's Just Mathematics—The Interpretation Is Up To You

One should never lose sight of the fact that the mathematical probability theory underlying our analyses is just mathematics; it provides little insight without an interpretation in terms of what we seek to understand.

Also, because it is just mathematics, how one makes the interpretation is not perfectly constrained by fixed rules.

Whereas it is common to model our base-level probability space in terms of capturing the units on which we are going to make measurements (height, weight, etc.), we could just as well start at a different level.

Example 4.20

Rather than letting $(\Omega, \mathcal{A}, P)$ model the population of interest as a set of individuals, I could just as well consider the outcome space to be {woman, man} and the equal probability sample of the population will produce one or the other outcome, with properly associated probabilities. Of course, the disadvantage of this is that we may find it difficult to understand the probabilities, as they are not derived from the simple assignment of $1/N$; nonetheless, one would not be sent to jail for taking this approach. Of course, the only sensible random variable would be an indicator variable (say, woman = 1 and man = 0), whereas using the previous model I can consider any possible measurement on humans—a considerably richer model.

Although modeling the data generating process often means specifying the objects of measurement as the outcome set, modeling subjective beliefs can take the more abstract approach.

Example 4.21

If I wished to model your confidence in the average height of people in the United States, I would specify my outcome set as the positive real line and determine a probability measure that reflects your confidence that the average height is in the various intervals in some appropriate sigma-algebra. Perhaps I use $(\mathbb{R}^+, \{\mathbb{R}^+, \emptyset, \{0 \text{ to } 6 \text{ ft}\}, \{\text{greater than } 6 \text{ ft}\}\}, P)$ as my probability space and P assigns your confidence judgments that the average height is less than 6 ft and greater than 6 ft.

Example 4.22

Suppose I am a frequentist wishing to investigate failure times in randomly selected individuals. If I consider that each person has a fixed failure time, then I might simply model the data generating process on the population of people as done above and consider failure time as a random variable. On the other hand, if I consider that nature assigns failure times for each individual by some random process, and as such each individual may have any possible failure time, then I might model my outcome set as the product $(\Omega \times \mathbb{R}^+)$. In other words, the data generating process combines my process of randomly selecting a person and nature's random assignment of failure time. The corresponding random variable is the observed failure time. The subtlety here is that the outcome is (w, t), for which many possible t's could have shown up even with the same w. The random variable is $X((w, t)) = t$.

The next section of the book provides examples of various uses and interpretations for probability models—some common, some not.

Additional Readings

Billingsley's book, titled *Probability and Measure* (Wiley Interscience, 1995), is an excellent text on probability theory presented in terms of measure theory. Resnick's book, titled *A Probability Path* (Birkhauser, 1999), is another such book.

For those interested in the topic from the disciplinary perspective of econometrics, Dhrymes' book *Topics in Advanced Econometrics: Probability Foundations* is worth reading. However, across both statistics and econometrics, my personal favorite is Davidson's book *Stochastic Limit Theory: Advanced Texts in Econometrics*—in my opinion, one of the best works on the subject.

Moving from probability to statistics, Schervish's book *Theory of Statistics* (Springer, 1995) provides an introduction to statistics in a measure-theoretic framework. It presupposes an understanding of measure theory and probability; however, it provides introductions to the topics in appendices.

Section III

Applications

5

Basic Models

The preceding chapters presented a conceptual introduction to measure-theoretic mathematical probability. This chapter will introduce uses of probability spaces as models for common research designs. We will start with the conceptually simple (i.e., modeling measurement error) and progress to a study design that may defy the usefulness of probability spaces as a modeling paradigm (i.e., modeling natural data generating processes for observational data).

Experiments with Measurement Error Only

Perhaps the simplest scientific use of a probability space is to model the variability associated with measurement in the context of a fully controlled experiment of a deterministic process. Consider an experiment in which all conditions are fully controlled except for chance error in the measurement instrument itself. In other words, for each run of the experiment, the physical outcome is the same, but the measurements may vary by chance. If we run the experiment N times, we record N measurements that may differ due solely to measurement error. How do we use this set of data to draw inferences regarding the underlying process?

We can model the measurement process for the experimental setup as a probability space: let Ω denote an outcome set comprising possible states that the measurement instrument can take, let $\mathcal{A}$ be an appropriately rich sigma-algebra, and let P denote a probability measure that models the objective uncertainty associated with the state of a given measurement outcome being in the sets of $\mathcal{A}$. Then $(\Omega, \mathcal{A}, P)$ is a probability space representing the uncertainty of the measurement process of the experiment. Suppose we define a function, labeling it Y, from each state of the measurement instrument to numbers on the real line. Now, if $\mathcal{B}$ is a sufficiently rich sigma-algebra on $\mathbb{R}$ such that the range of the function $Y^{-1}(B)$ for all $B \in \mathcal{B}$ is contained in $\mathcal{A}$, then $(\mathbb{R}, \mathcal{B})$ is a measurable space and Y is measurable $\mathcal{A}$: Y is therefore a random variable representing variation in the measurement process. If we let $p(B) = P(Y^{-1}(B))$ define a measure on $(\mathbb{R}, \mathcal{B})$, then $(\mathbb{R}, \mathcal{B}, p)$ is a probability space associated with the random variable Y. Characteristics of p provide information useful for making inferences about the experiment accounting for measurement error.

For N independent runs of the experiment, there is one realization from each of N random variables from the probability space. If the experimental runs are identical, it is reasonable to presume that each of these random variables has the same distribution, and we can use statistics (a function of the N random variables, which consequently is itself a random variable) to estimate properties of p and thereby provide information for making inferences. For example, the histogram of the data reflects p from the N realizations from the probability space. If we are interested in the expected value of Y taken with respect to p as a summary of the experimental outcome, then the sample mean and confidence interval provide a reasonable estimate of this quantity, as shown in any introductory statistics text. If our interest is not in the estimate itself, but instead we are interested in testing a particular hypothesis, then we can proceed with statistical hypothesis testing. For example, suppose we have a theory that implies a state corresponding to a perfect measurement of less than 10. If we are correct in assuming that the measurement instrument provides unbiased results, then the mean of the distribution for Y with respect to p will also be less than 10. We record the measurements for N independent runs of the experiment and calculate a sample mean of 9. This result conforms to our prediction, but can we rule out the alternative explanation that the true value is 10 or greater and we got a sample mean of 9 by virtue of measurement error alone? Here we have a classic setup for a statistical test to rule out an alternative explanation for the data.

Experiments with Fixed Units and Random Assignment

Another use of probability spaces is to model uncertainty associated with random assignment of research subjects to experimental conditions. In this setup the probability space is not used either to model uncertainty regarding which research subjects are used (they are considered fixed) or to model measurement error, which is considered to be zero. Instead, the probability measure is defined to represent the uncertainty that a subject is assigned to each possible experimental condition. In this case, for each subject, we may define an outcome space of $\Omega = \{C_1, C_2, \ldots, C_K\}$, in which each element is a particular experimental condition with a corresponding sigma-algebra $\mathcal{A}$ as the power set of Ω. For an experiment with N subjects, $(\Omega^N, \mathcal{A}^N)$ is a measurable space representing the possible configuration of experimental conditions assigned across the group of subjects. For P, a probability measure representing the probability of assignment to conditions for the N subjects, $(\Omega^N, \mathcal{A}^N, P)$ is the probability space that models assignment.

Unfortunately, using $(\Omega^N, \mathcal{A}^N, P)$ as our model is restrictive in its representational benefits. The outcome space Ω^N contains information regarding experimental conditions but does not have individual-specific information. In other words, we can define random variables that indicate what condition was assigned, but it is not apparent how outcomes are available for each individual in each condition. Consider only one subject (i.e., $N = 1$) with one treatment (T) and a control (C) assignment possible: the probability space is $(\{T, C\}, \mathcal{A}, P)$. A random variable is an $\mathcal{A}$-measurable function of $\{T, C\}$. It is not clear how we obtain the person-specific measures based on T or C alone: Suppose T is a particular medication and we are interested in a subject's blood pressure. Our random variable assigns a number to the condition of taking the medication and a number to the condition of not taking the medication. However, do such conditions have a blood pressure? No. Neither medications nor the condition of taking medications is the kind of things that can have a blood pressure. Of course, we can simply declare the function and, making its assignment correspond to a particular subject's measurements, multiple subjects would then require multiple functions; however, this makes the model somewhat opaque. So, for didactic purposes, an alternative would be to consider our outcome space to comprise the set of all qualities of the subject (say subject s) under condition T and the set of all qualities under the condition C: $\Omega = \{\{X_s\}_T, \{X_s\}_C\}$. As before, there are only two elements to this outcome set: one element is the set of characteristics associated with subject s under condition T, and the other is the set under condition C. Meaningful random variables are now more evident: some obvious ones include measurements on the real line of the X qualities as well as treatment assignment.

Sticking to our one-subject experiment with only two treatment conditions, suppose we model the experiment with $(\Omega, \mathcal{A}, P)$ for $\Omega = \{\omega_1, \omega_2\}$ with ω_k being either the simple indicator of treatment and control or the sets of qualities under treatment and control conditions. For an appropriately defined random variable Y, the parameters of the distribution of Y are readily calculated. For example, the expected value of Y is $Y(\omega_1) \cdot P(\{\omega_1\}) + Y(\omega_2) \cdot P(\{\omega_2\})$. For those qualities that do not change, $Y(\omega_1) = Y(\omega_2)$, and the expectation is simply the value; for those qualities that differ across treatment groups, the randomization-weighted average becomes the expectation.

For example, let $\Omega = \{\{\text{male, blood pressure, } T\}, \{\text{male, blood pressure, } C\}\}$, in which case the outcome space is the sex and blood pressure state of the individual under treatment condition T and control condition C. Define $\mathcal{A}$ to be the power set of Ω, and P is 0.5 for each element of the most granular partition of Ω in $\mathcal{A}$ (i.e., equal probabilities of being assigned to treatment or control). Now let us define three random variables: $Y = 1$ if male, 0 otherwise; $X = $ blood pressure measurement (suppose it is 130 under the treatment condition and 150 under the control condition); and $Z = 1$ if treatment condition, 0 otherwise. The expected value of Y is $1 \times 0.5 + 1 \times 0.5 = 1$, and the variance of Y is $(1 - 1)^2 \times 0.5 + (1 - 1)^2 \times 0.5 = 0$. The expected value of X is $130 \times 0.5 + 150 \times 0.5 = 140$, and the variance of X is $(130 - 140)^2 \times 0.5 + (150 - 140)^2 \times 0.5 = 100$. The expected value of Z is $1 \times 0.5 + 0 \times 0.5 = 0.5$, and the variance of Z is $(1 - 0.5)^2 \times 0.5 + (0 - 0.5)^2 \times 0.5 = 0.25$. The treatment effect on sex is $E(Y \mid Z = 1) - E(Y \mid Z = 0)$, which is 0 (i.e., sex does not change), whereas the treatment effect on blood pressure is $E(X \mid Z = 1) - E(X \mid Z = 0)$, which is -20.

Of course, knowledge of the variable values associated with each outcome in the outcome space is typically unknown except for the few outcomes that are measured. Consequently, multiple experiments with random variables having the same distribution (or shared parameters of interest) are typically used so that statistics are available to estimate desired quantities.

Observational Studies with Random Samples

A common study design among applied researchers is to collect a sample from a population and measure their characteristics. For example, perhaps we wish to know the distribution of blood pressure among the adult population in the United States. We might propose to measure the blood pressure of all adults in the population, but that would likely be impractical. So, instead we might propose to collect a sample of adults from the population and measure their blood pressures. However, now the set of blood pressures we obtain will likely vary depending on the sample we happen to get, and the sample we happen to get will likely vary due to how we go about obtaining the sample. Consequently, there may be variation in the observed numbers (e.g., sample averages and variances) due to the mechanism of sampling, that is, due to the data generating process. How can we account for this variation? One strategy is to base our analysis on a probability space that models the data generating process producing this variation.

To proceed with this example, suppose we have a sampling frame of all people in the United States (e.g., a list of unique identifiers and viable contact information). Suppose we plan to randomly select a sample of individuals from the frame such that each person has an equal chance of being selected. For simplicity in this example, let us further suppose we plan to sample with replacement (i.e., after selecting a name, we put the name back so it could possibly be selected again). Moreover, again for simplicity, suppose that we can compel the measurement of all those whom we select and that measurements are precise. We can model this process by letting the outcome set be the population of the United States and the associated event set be the power set of the outcome set; we define the probability measure based on the sampling probability associated with each unit set in the event set (i.e., each individual in the population):

Ω is the population of people in the United States.

$\mathcal{A}$ is the power set of Ω, allowing for probabilities to be assigned to any subset.

P is based on the sampling probabilities of the individuals w in the population, $w \in \Omega$, that is, $P(\{w\}) = 1/N$ for all unit sets $\{w\}$ in $\mathcal{A}$, and N denotes the number of people in the population.

By these definitions, $(\Omega, \mathcal{A}, P)$ provides a model of the data generating process that will allow us to account for variation in sampling. If we define $Y(w)$ to be a real-valued function on Ω to reflect a measure of systolic blood pressure, then for Y measurable $\mathcal{A}$, we have $(\mathbb{R}, \mathcal{B}, p)$, with $p(B) = P(Y^{-1}(B))$ for all $B \in \mathcal{B}$, as a probability space that defines the distribution of the random variable representing blood pressure measurements, and we can use our usual statistical methods to estimate parameters of this distribution or test hypotheses.

If instead we generate the data by taking a random sample of states (with replacement, for simplicity) and then we take a random sample of individuals from within each state (again, with replacement), we would have a simple cluster (or nested) sampling design. The outcome set of this data generating process may be defined as the product set of states S and US population Ω (i.e., $S \times \Omega$), the event set is again the power set of the outcome set, and P represents the cluster data generating process. For example, P is such that $P(\{s, w\}) = P_S(\{s\}) \cdot P_{s \times \Omega}(\{w\} \mid \{s\})$, which is the probability of obtaining a state s in the first step of the data generating process multiplied by the conditional probability of obtaining individual w given one has obtained state s in the first step of the data generating process. The random variable Y then reflects the distribution of blood pressure associated with P, which in this case is the distribution associated with a cluster sampling design.

Consider a third strategy. Suppose we engage the strategy of the first example, but in this case we cannot compel participation in our study: each individual may or may not agree to participate. In this case, our outcome set may be modeled as $\Omega \times$ {agree to participate, not agree to participate}, our event space is again the power set associated with this outcome set, and our probability is defined as $P(\{w, a\}) = P_\Omega(\{w\}) \cdot P_{w \times A}(\{a\})$, the probability of selecting an individual w multiplied by the conditional probability of agreeing to participate given that the individual is selected. The random variable for blood pressure now has a distribution reflecting the combination of how we select individuals and whether those individuals are inclined to participate in our study. If those with high blood pressure are less inclined to participate, then the distribution of y will reflect this by shifting toward lower values of blood pressure, and vice versa if those with high blood pressure are more inclined to participate.

These examples highlight an important point: Statistical analyses provide estimates or tests of the parameters associated with random variables, and these random variables derive their distributions from the underlying probability space being used. Consequently, when modeling different data generating processes that are implemented on the same population, there may be different distributions for the same type of measurement (e.g., blood pressure). Using an unbiased estimator for each random variable will provide unbiased information about the parameters related to the data generating process, but it may not necessarily correspond to the underlying population. If you wish to estimate the average blood pressure in the population, but you use the third data generating strategy (i.e., the one with self-selection), then the unbiased estimator of the mean for the random variable Y may not be unbiased for the population average itself (perhaps only those with low blood pressure are willing to participate). Again, your statistics are estimating parameters of a probability space modeling a particular data generating process, which may not correspond to the population-level information that you seek. Consequently, it is necessary to design the data generating process and analytic strategy so that functions of the related parameters correspond to the population quantities of interest.

Experiments with Random Samples and Assignment

If we collect a random sample of individuals and then randomly assign each individual to one of a set of conditions, then our model of the data generating process is based on the product outcome space of both the random assignment and random sample models considered separately above. For example, suppose we collect a sample of individuals from a population Ω and assign each individual to a treatment or control condition from the set T = {treatment, control}. We can model this data generating process using the outcome set specified as $\Omega^* = \Omega \times T$. With an appropriately defined sigma-algebra, the probability can be written as the product of the marginal probability of selecting an individual from Ω and the conditional probability of treatment given the individual selected: $P(\{w, t\}) = P_\Omega(\{w\}) \cdot P_{w \times T}(\{t\})$. If treatment assignment is independent of individual selection, then as we know this probability reduces to the product of the two marginal probabilities: $P(\{w, t\}) = P_\Omega(\{w\}) \cdot P_T(\{t\})$.

We should keep in mind that the structure of $P(\{w, t\})$ (i.e., whether it reduces to the product of marginals or not) for each observation does not bear on our judgment of whether *observations* are independent or not. Such dependence is a question to be asked of the bicoordinate probabilities associated with the higher-order measurable space $(\Omega^{*N}, \mathcal{A}^N)$ reflecting the data generating process underlying the selection of a sample of N outcomes. As presented in preceding sections, this case is one in which observations are independent, and consequently the random variables across observations are independent as well.

Suppose, instead, that our data generating process is based on cluster sampling with random assignment of conditions. Perhaps we take a random sample of elementary schools (with replacement) from the population of elementary schools in the United States, we then take a sample of children (with replacements) from each school's population of students, and finally we randomly assign each student to an experimental condition. Our outcome set is made up of the US population of elementary schools (S), the US population of elementary schoolchildren (C), and the set of experimental conditions (T): $\Omega = S \times C \times T$. Our event set is the sigma-algebra generated by the product of the sigma-algebras defined for each component of the outcome set: $\mathcal{A} = \mathcal{A}_S \otimes \mathcal{A}_C \otimes \mathcal{A}_T$. Moreover, our probability is constructed to model the cluster sampling and random assignment so that our random variables will reflect this data generating process and our corresponding standard errors will represent variation in the estimators due to the sampling uncertainty of the data generating process: $P = P_S \cdot P_{C|S} \cdot P_{T|C,S}$, which is the product of the probability of getting a particular school with the probability of obtaining a particular child within that school and the probability of assigning a particular treatment to that child. If treatment assignment is random without regard to school or child, then P reduces to $P_S \cdot P_{C|S} \cdot P_T$.

Again, dependence between observations is identified by considering the bicoordinate probabilities associated with the higher-order measurable space $(\Omega^N, \mathcal{A}^N)$. As preceding sections have shown, for cluster samples, observations within clusters are dependent and observations across clusters are independent (when equal probability sampling is done with replacement). Random variables for observations across clusters are independent. By contrast, dependence of random variables for observations within clusters will depend on whether the distribution of the random variables is the same across all clusters: if so, the corresponding random variables are independent; if not, they are dependent for observations within a cluster—in this case, within a school.

Would the situation change if we engage a cluster randomized trial in which we obtain a sample of schools, randomly assign the school to an experimental condition, and randomly select a sample of students within each of the selected schools? All students within a given school will be subject to the same experimental treatment. In this case, the probability space representing each observation would be similar to that presented in the preceding paragraph.

Table 5.1 presents the sampling probabilities of particular school (S), experimental condition (T), and child (C) combinations. Remember, the probabilities in the table sum to 1.

TABLE 5.1

Probabilities for First Observation of Cluster Randomized Trial

School	Experimental Condition	Children							
		C_1	C_2	C_3	C_4	C_5	C_6	C_7	C_8
S_1	T_0	P_{101}	P_{102}	P_{103}	P_{104}	0	0	0	0
	T_1	P_{111}	P_{112}	P_{113}	P_{114}	0	0	0	0
S_2	T_0	0	0	0	0	P_{205}	P_{206}	P_{207}	P_{208}
	T_1	0	0	0	0	P_{215}	P_{216}	P_{217}	P_{218}

TABLE 5.2

Probabilities for Second Observation of Cluster Randomized Trial given that First Observation is (S_1, T_0, C_3)

School	Experimental Condition	Children							
		C_1	C_2	C_3	C_4	C_5	C_6	C_7	C_8
S_1	T_0	p_{101}	p_{102}	p_{103}	p_{104}	0	0	0	0
	T_1	0	0	0	0	0	0	0	0
S_2	T_0	0	0	0	0	0	0	0	0
	T_1	0	0	0	0	0	0	0	0

TABLE 5.3

Probabilities for Second Observation of Cluster Randomized Trial given that First Observation is (S_2, T_1, C_5)

School	Experimental Condition	Children							
		C_1	C_2	C_3	C_4	C_5	C_6	C_7	C_8
S_1	T_0	0	0	0	0	0	0	0	0
	T_1	0	0	0	0	0	0	0	0
S_2	T_0	0	0	0	0	0	0	0	0
	T_1	0	0	0	0	p_{215}	p_{216}	p_{217}	p_{218}

Given that we obtained (S_1, T_0, C_3) on one observation, the conditional probability of another observation within the same cluster becomes as represented in Table 5.2, and again the probabilities sum to 1.

By contrast, the probabilities associated with the second observation given the first observation obtained (S_2, T_1, C_5) are presented in Table 5.3 in which the probabilities sum to 1.

In Tables 5.2 and 5.3, the probability distributions vary by which school and condition was obtained in the first observation: the observations are dependent within school and condition. Of course, since each school is assigned only one treatment condition, this reduces to dependence within school. Similar to the usual cluster sample design, observations within cluster are dependent, whereas observations across clusters are independent. The corresponding random variables are dependent if their distributions vary across school and condition.

Observational Studies with Natural Data Sets

Another common source of data that requires careful consideration is what I call a *natural data generating process* (NDGP): this is a data generating process that is not designed but is presumed to underlie data that are simply collected as they occur. Examples include data from all patients that visited a clinic in some time period; New York's Statewide Planning and Research Cooperative System, which collects patient-level detail on patient character-istics, diagnoses and treatments, services, and charges for every hospital dis-charge, ambulatory surgery patient, and emergency department admission in New York State; and the National Cancer Institute's Surveillance, Epi-demiology, and End Results registries, which collect data on each cancer patient in participating regions of the United States.

The distinguishing feature of such data sets is that they result from meas-urements on individuals who were not selected by virtue of some designed data generating process but rather who happen to "show up" by virtue of some unknown natural process and be recorded: those individuals in a pop-ulation who happen to seek care from a given clinic at a given time; those individuals in New York who happen to be admitted to a hospital; those individuals in certain US regions who present with cancer to the medical sys-tem. What at first may seem like a trivial question is in fact a perplexing problem: how do we model an NDGP? This is a question that may not have a satisfactory answer, leaving us with a suspicious interpretation of analyses. Examples 4.9 and 4.10 in Chapter 4 presented two simple scenarios with NDGPs. Let's consider some other examples.

Suppose we are interested in modeling the natural probability that a per-son gets some specified illness in a given year and that we have access to data on each person in a population regarding whether that person got this illness. We are not asking about the proportion of people who actually got sick in that year. Instead, we presume that those who happened to get sick could well not have gotten sick, and those who did not get sick could well have. Our interest is in modeling the natural uncertainty about an indi-vidual getting this illness and estimating a corresponding probability parameter.

We could begin by specifying the population as the outcome set, the power set as the event set, and a probability modeling nature's selecting of individ-uals. For a large population, this probability space might be a reasonable approximation, but the specification is flawed in its particulars. Suppose our outcome set is the population, our probability measure represents an NDGP that assigns a probability to each person in the population of getting sick, and nature engages this NDGP a number of times to generate the peo-ple who get sick. However, what is the probability that someone will get sick each time the NDGP is engaged? By the definition of a probability space, it is one: someone must get sick each time the NDGP is engaged. Moreover, only one person gets sick each time the NDGP is engaged; therefore, we need to model a process by which a number of people may get sick. Because the number of people who get sick is uncertain, we need to augment the sick-generating NDGP with a number-of-people-generating NDGP in which nature picks the number N of people who will get sick and thereby the number of times the sick-generating NDGP is engaged. If N is 0, then the sick-generating NDGP is never engaged; if N is 1, then it is engaged once, and so on. If not this strategy, then some other clever organization of probability models that serve the same function must be developed (e.g., using the power set of the population as the outcome set would allow for nature's selecting any number of people to be sick).

An alternative is to model each person's natural uncertainty separately. In a simplified form, perhaps the outcome space for each person is defined as S = {not sick, sick} with corresponding algebra $\mathcal{A}$ = {S, $\varnothing$, {not sick}, {sick}} providing us with a measurable space on which each person w has a probability p_w defined that reflects their natural susceptibility to getting sick. Consequently, for each person w in population Ω, we have a probability space $(S, \mathcal{A}, p_w)$. The product space across Ω provides the overall probability space reflecting the population: $(S^N, \mathcal{A}^N, P)$, in which P reflects the combination of the p_w's across the individuals in population Ω. Now that each person may or may not get sick, it is possible for any number of people to get sick, or no one at all.

Unfortunately, without additional constraints, this specification is not amenable to analysis. If each person has an idiosyncratic probability of getting sick, then there are N probability parameters and N observations. If, by contrast, it is reasonable to model the probability of an arbitrary person getting sick as a function of specific characteristics and a small set of fixed common parameters, we can use the information across the whole population to estimate those parameters.

As a simple example, suppose $p_w = p$ for all $w \in \Omega$ (i.e., everyone has the same probability of getting sick). Moreover, suppose this illness is such that each person's probability is independent of whether others get sick. Then $(S^N, \mathcal{A}^N, P)$ is a probability space with P equal to a product measure comprising individual probabilities p of getting sick and $(1 - p)$ of not getting sick. Because the data comprise realizations of random variables with a common p, we can estimate p from the data. For example, by maximum likelihood our estimator of p is $\hat{p} = N_s/N$, in which N_s is the number of people in the population who got the specified illness. The variance of our estimator is $\mathrm{var}(\hat{p}) = (p \cdot (1 - p))/N$. Note that the usual "large sample" asymptotic properties of maximum likelihood estimation hold in terms of population size rather than the usual consideration of sample size.

Contrast this with the goal of estimating the proportion of people who got sick in the population. We then calculate the proportion as $\pi = N_s/N$, which is the same quantity that our estimator $\hat{p}$ would produce as an estimate, except that in the case of π, we do not have an estimate of a data generating process parameter. Instead we have the actual proportion—a population parameter. Whereas the estimator $\hat{p}$ has a variance associated with possible realizations of which people could become sick in the population, π does not have such a variance associated with it. By one account we can say that the proportion of people who got sick was π; by another account we can say the probability of a person getting sick is estimated as $\hat{p}$, which will have a variance of $\text{var}(\hat{p})$. The difference between these two accounts derives from what is being modeled, not from the data.

Now let's consider we have an equal probability random sample (generated with replacement) of n people from the population. The estimator $\tilde{p} = n_s/n$, in which n_s is the number of people in the sample who were sick, provides an estimate of a parameter for the sampling data generating process, which corresponds to π and has a variance $\text{var}(\tilde{p}) = (\pi \cdot (1 - \pi))/n$. On the other hand, if we consider $\tilde{p}$ as an estimator of the probability of a person getting sick, then, unfortunately, it is likely to be biased due to the difference between the equal probability sampling and the unequal probabilities of getting sick (an interesting mirror image of classic self-selection bias, which would require similar strategies to handle).

Our assumption that $p_w = p$ for all $w \in \Omega$ does not at all seem plausible for any real illness. After all, how likely is it that each individual has the same probability of getting a particular illness? If this assumption is sufficiently in error, then $\hat{p}$ becomes meaningless as an estimator of an individual's probability of getting sick, whereas both π and $\tilde{p}$ remain meaningful in their original interpretations. To be clear, in this case $\hat{p} = N_s/N$ makes no sense but $\pi = N_s/N$ does. How can two quantities that are the same differ in being meaningful or not? They do so because they differ in meaning: $\hat{p}$ estimates a parameter of an NDGP that represents the natural probability of a person getting sick, whereas π is the proportion of people who got sick. Similarly, $\tilde{p}$ is an estimator of a sampling data generating process, which, if the data generating process is indeed as we specified, is also consistent for π and retains its meaning accordingly.

Suppose we have a vector of characteristics x for each individual in the population such that the natural probability of an arbitrary individual w getting sick can be reasonably modeled as a function g of characteristics x and corresponding parameter θ: $p_w = g(x_w, \theta)$. In other words, probabilities of getting sick vary across individuals in the population because individuals have different characteristics as measured by x. Note that by our current specification of (S^N, A^N, P), our only random variable is an indicator of illness. So, what is x? Well, it is not a random variable, but it is an individual-specific observed parameter that specifies the individual probabilities. Now the probability space (S^N, A^N, P) can be the basis for estimating the common unobserved parameter θ, which in combination with an individual's specific x would provide us with an estimate of that person's probability of getting sick. Moreover, the point and interval estimators of θ will give us information about how these natural probabilities of getting sick are related to the idiosyncratic parameters x.

In the preceding example, I assumed observations were independent. However, how might we determine whether observations ought to be modeled as dependent? Determining whether two arbitrary individuals i and j are dependent requires considering the bicoordinate subspace $(S_i \times S_j, A_i \otimes A_j, p)$.

We can represent the meaningful components of the Cartesian product of events as a simple two-by-two table:

$$
\begin{array}{c c|c|c|}
 & & \multicolumn{2}{c}{j} \\
 & & s & ns \\
\hline
 & s & p_{s,s} & p_{s,ns} \\
\cline{2-4}
i & ns & p_{ns,s} & p_{ns,ns} \\
\hline
\end{array}
$$

Does the probability of illness for individual j given that individual i is ill differ from the probability of illness for individual j given that individual i is not ill? The answer depends on the data generating processes.

Assume that we are interested in an illness stemming from exposure to an environmental toxin. It might be reasonable to presume that, although P_i and P_j may both be high or low for individuals with similar exposures, the probability of illness for one individual is the same regardless of whether the other was selected by nature to be ill; therefore, the probabilities are independent. As an aside, consider how our judgment in this regard might change if we were modeling subjective beliefs about how likely these people were to get ill rather than the frequentist approach taken here.

What if we were interested in an infectious disease? Now we might consider that these two observations are dependent if the process by which nature assigns illness to one also impacts the other. In a fully connected society (i.e., there is a pathway of individuals that connects everyone in the population), we might consider every pair of observations as dependent, with the degree of dependence a function of how many pathways connect the individual's. However, identifying that the state of illness is dependent between all pairs of individuals is not of much help because we only have one observation per person. We need to model the dependence if it is to be useful in our analysis. Perhaps categorizing by family or neighborhood would work, or perhaps using some distance measure.

The preceding example focused on having data on the full population; however, it is more common that we have data on only some of the population, as mentioned in the introduction to this section. To illuminate issues regarding modeling an NDGP in this case, let's revisit the examples provided in Chapter 4.

Consider that you have all the study-relevant patient data for a specific day of a health care clinic. The physician who attended in the morning did a better job at treating the outcome of interest than the physician who attended in the evening. How do we determine whether the observations of the NDGP are dependent?

If nature independently samples N people with replacement from a population Ω and independently samples a time of the day, with replacement, from a set T, for each person and then sends the N people to the clinic at their designated times, then it should be clear that the observations are independent. The probability of a person–time pair of one observation does not depend on the actual person–time event occurring on any other observation. However, if nature does a sequential sampling strategy by which people are sent to the clinic in the order of the sequential sample, then the probabilities associated with events on the outcome set $(\Omega \times T)$ changes with the sequence. For example, if the first observation has a time of 10 am, then the probabilities associated with events on the outcome set $(\Omega \times T)$ for the second observation must assign zero to all event with T less than 10 am. The third observation is constrained by the previous, and so on. The observations are dependent. Are the outcome variables dependent? Because physician shift is fixed according to time of day (morning/evening) and the morning physician does better than the evening one on a particular outcome, the distribution of outcomes changes with each observation in the sequence indicating dependence in the outcome variables. Note, as I've mentioned at various points in this chapter, that the data could well be the same for both processes and therefore cannot differentiate the models in this case.

Population Models

The preceding models use probability spaces to represent sources of uncertainty. However, remember that mathematical probability is not substantively interpreted in itself and is therefore not restricted to modeling uncertainty. In fact, the main targets of investigation are often in terms of a probability model having a different interpretation.

As researchers, we are typically not interested in the parameters of a data generating process (except perhaps in the case of a natural data generating process); a random variable from a sampling process is seldom inherently interesting. Most likely, we are using sample information about a parameter of the data generating process to provide information about a target population. We are interested in parameters of a population model.

A *population model* is a probability space defined to represent the normalized histogram of variables and their relationships in a given population. In this case, the outcome space (Ω) is defined as the population of interest, the event space is the power set of the outcome space $(A = \wp(\Omega))$, and the probability (P) is defined such that $P(\{w\}) = 1/N$ for all elements w in Ω, in which N denotes the total number of elements in Ω. It is important to note that the probability is not modeling a data generating process: the $1/N$ does not denote an equal probability sampling strategy. Instead, it defines a weight that allows us to represent the associated random variables according to their actual normalized frequencies and relations in the population. The probability space (Ω, A, P) is then a population model, and the parameters of associated distributions for random variables on this probability space are called *population parameters*.

Suppose we were interested in the relationship between body mass index and systolic blood pressure in the United States. We can use a population model (Ω, $\mathcal{A}$, P) in which Ω represents the population of people in the United States, and the other components of the probability space are defined as above. Let measures of body mass index (BMI) and systolic blood pressure (BP) be random variables that are measurable with respect to $\mathcal{A}$. Then the joint distribution of BMI and BP in the population derives from P, which is defined to assure that the joint distribution reflects the normalized histogram of BMI and BP across the population. Consequently, the parameters of this joint distribution are population parameters that represent the actual distribution of BMI and BP as they exist in the population. For example, the regression $E(BP \mid BMI \in b)$ (i.e., the expected value of BP given that BMI has a value in some small interval of values denoted by b) reflects the average BP in the subpopulations having the specified BMI values.

Note that, although the random variables are likely to have standard deviations, there is no meaningful sense of a standard error associated with random variables of a population model because these models are not representing data generating processes. If you could measure the BMI and BP for all persons in the population of interest, then the parameters of their distribution in the population may be directly evaluated—there is no sampling error and consequently no standard errors (which are data generating process concepts). See Chapter 6 for further discussion of standard errors.

Data Models

Probability measures can also be used to model data, but here we must be careful not to confuse a model of data with a model of a data generating process. Whereas a model of a data generating process represents uncertainty regarding the possible outcomes the process could obtain, a model of data does not represent such uncertainty, because the outcome set represents the data in hand rather than a set of possible outcomes.

Suppose we define an outcome set as comprising the data elements in a table of data and the corresponding event set as the power set. Leaving aside the question of what definition of a probability measure would be useful, we immediately run into the question of what useful random variable could be defined. Because the elements on the outcome set are numbers, presumably the random variables would be functions of those numbers. However, by our specification, the outcome set is an undifferentiated set of numbers (no longer labeled by variable names, etc.); consequently, it is unclear what purpose such functions would serve.

It can be better to partition the data by the objects being represented. For example, it is common for rows of a data table to represent measurements

TABLE 5.4

Representation of a Data Set in Terms of Rows and Columns

	Height	Weight	Blood Pressure	Cholesterol
Row 1	66	150	120	190
Row 2	60	155	143	225
⋮	⋮	⋮	⋮	⋮
Row n	72	210	138	250

on specific objects (e.g., people or hospitals): each row is a vector of measures for a different object (e.g., blood pressure or number of beds). In this case, we can model data in an indirect fashion by defining the outcome set as the designated partition. For example, the set of rows in the data set: $\Omega = \{r_1, \ldots, r_n\}$ for n rows (e.g., see data in Table 5.4). Now we can define random variables as functions of the specific elements of the vectors (rows in the table), labeling them according to their meaning (blood pressure, age, etc.). Our probability space is then $(\Omega, \mathcal{A}, P)$, in which the set Ω is the data partition of interest, the sigma-algebra $\mathcal{A}$ is sufficiently rich for our purpose (e.g., the power set), and P reflects a useful probability measure. The random variables will then have corresponding distributions reflecting data that preserve a useful interpretation, being functions of specified measurements (e.g., again, blood pressure and age) across the partition representing the objects that were measured.

What would be a useful definition of P? Perhaps the most useful and common definition is to set $P(\{r\})$ equal to $1/n$ for all n elements r of the outcome set Ω, which, for example, may be the rows of a data table. This specification will allow our random variables to represent the normalized histogram of measures in the data across the partition. In fact, by defining our probability space this way, we are treating the set of objects (i.e., the elements of a partition) on which we have measurements (data) as its own population and are defining a corresponding population model as defined in the preceding section. This data model is the basis for the sample descriptive statistics.

Another useful definition of P is as renormalized sampling probabilities of a data generating process that produced the data. Such a data model allows for random variables that reflect an underlying population from which a nonequal probability data generating process obtained the sample. Herein lies the basis for probability-weighted statistics of a data model.

Connecting Population and Data Generating Process Models

If we do not have access to measures on all members of the population, and therefore cannot directly calculate parameters of a population model, we may resort to defining a data generating process by which we can collect a sample to inform our questions. Suppose we are interested in the parameters of a population model $(\Omega, \mathcal{A}, P)$ in which $P(\{w\}) = 1/N$ for all w in Ω to facilitate describing the normalized histogram of characteristics in the population. We do not have access to the full population, but we can obtain data from an equal probability sample with replacement. The measure space for each observation of the data generating process is $(\Omega, \mathcal{A}, p)$ in which $p(\{w\}) = 1/N$ for all w in Ω. Because p, which models a data generating process, assigns the same probabilities to each set in $\mathcal{A}$ as does P, which is instrumentally defined for a population model to be $1/N$ for all unit sets, then the distributions for all random variables of the data generating process model are the same as those for the random variables of the population model. Collecting information about the data generating process model can inform the population model.

If instead p is a model of sampling with other than equal probabilities, then we can create random variables associated with $(\Omega, \mathcal{A}, p)$ that inform population parameters. For example, suppose we define a measurable function X on $(\Omega, \mathcal{A})$, and we are interested in its mean with respect to $(\Omega, \mathcal{A}, P)$, the population model—that is, we want the mean of X in the population. We have a data generating process that produces X as a random variable on $(\Omega, \mathcal{A}, p)$ for which p does not assign equal probabilities to each possible person in the population such that the distributions of X related to P and p are not the same. If we define a random variable as follows:

$$Z(w) = X(w) \cdot \frac{P(\{w\})}{p(\{w\})} \tag{5.1}$$

then the expected value of Z with respect to p is equal to

$$E_p(Z) = \sum_\Omega \left(\left(X(w) \cdot \frac{P(\{w\})}{p(\{w\})} \right) \cdot p(\{w\}) \right) \tag{5.2}$$

but the probabilities p in Equation 5.2 cancel such that the summation reduces to the simpler summation

$$\sum_\Omega \left(\left(X(w) \cdot \frac{P(\{w\})}{p(\{w\})} \right) \cdot p(\{w\}) \right) = \sum_\Omega (X(w) \cdot P(\{w\})) \tag{5.3}$$

which is equal to the expected value of X with respect to the probability P:

$$\sum_\Omega (X(w) \cdot P(\{w\})) = E_P(X) \tag{5.4}$$

Weighting by a function of sampling probabilities provides a random variable with a mean equal to the population mean (actually, and more generally, P can be any probability and not merely one associated with a population model). Although statistical estimation is not the focus of this book, note that when P represents a population model, or more generally when $P(\{w\})$ is a constant across all $\{w\} \in \mathcal{A}$, P drops out of most estimators, leaving $1/p(\{w\})$ as the relevant factor, called the *sampling weight*.

Connecting Data Generating Process Models and Data Models

A fundamental connection between a model of a data generating process and a model of data is represented in Figure 5.1. The top section of Figure 5.1 indicates that we specified a model of a data generating processes as $(\Omega, \mathcal{A}, p)$, which will be used to specify n different random variables X_1 to X_n (representing the same type of measurement—e.g., blood pressure for each). Because each random variable reflects the same model of a data generating process and the same type of measurement, each of the n random variables has the same distribution F. One realization for each random variable is possible; consequently, we can obtain a single data point for each random variable: datum x_1 for random variable X_1, datum x_2 for random variable X_2, and so on. Put together, this set of data generating processes, random

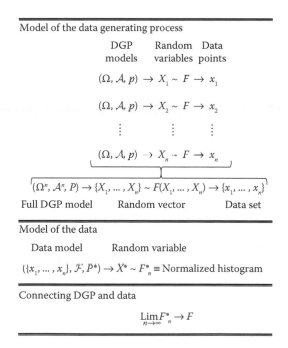

FIGURE 5.1
Connecting a data generating process model to a data model.

variables, distributions, and data compose the overall measure space (Ω^n, $\mathcal{A}^n$, P) with corresponding random vector $(X_1, X_2, \ldots, X_n)^T$, having a joint distribution $F(X_1, X_2, \ldots, X_n)$, and resulting dataset $\{x_1, x_2, \ldots, x_n\}$.

We can model the resulting data with the probability space $(\{x_1, x_2, \ldots, x_n\}$, $\mathcal{F}, P^*)$ as in the middle section of Figure 5.1. As a data model, the outcome set represents the data, $\mathcal{F}$ is the power set of the data, and P^* is equal to $1/n$ for all sets containing individual data elements representing the objects that were measured. P^* does not represent a data generating process. It is specified, as with common population models, to allow the distribution of X^* to represent the normalized histogram across the data. In the preceding general discussion of data models, it was better to define the outcome set as a partition of the data. By contrast, in this example, each data point represents the same quality (e.g., blood pressure) and thereby the meaning of each datum is not ambiguous, and the random variable of interest on this space will simply be the identity function $X^*(x_i) = x_i$. Thus, we can just as well use the actual set of data as the outcome space for Figure 5.1 without generating ambiguity. The random variable X^* has a distribution F_n^*, which represents the normalized histogram of the x_i in the data.

At this point the only evident connection between the data generating process model and the data model is merely the fact that the latter uses the data from the former as its outcome set. However, a more important connection exists. If the data are realizations of random variables having the same distribution, then as n gets large and continues to increase (formally, we'd say as n goes to infinity), the histogram of X^* (and therefore its distribution) will get closer to the histogram of each X (and therefore its distribution). This is formally stated as the limit, as n goes to infinity, of F_n^* is equal to F (or more precisely, F_n^* becomes arbitrarily close to F). In practical terms, this fact is the basis for asymptotic properties underlying classic bootstrap estimation in which samples are taken from the data as if the data constituted the distribution from which the original sample was taken.

To restate, one connection between a data generating process model and the model of its corresponding data is that as the sample size increases, the distribution in the data begins to look like the distribution associated with the data generating process. Therefore, resampling (i.e., bootstrapping) from large data sets can mimic the process underlying the original sample and provide a means of estimating the variation in an estimator due to the data generating process. Note that we have introduced the notion of sampling from the data itself, which can be modeled as another sampling process.

An extension of this connection between models of data generating processes and models of data is found in estimation by method of moments. The idea behind the method of moments estimator is to estimate parameters of interest by equating moments of a distribution from a data generating process with moments of the distribution from the data (called the *sample moments*) and to solve these equations for the data generating process parameters of interest. For the simplest of examples, suppose the mean of a random variable X from a data generating process is μ, and the mean of the corresponding X^* from the data is $\overline{X} = \frac{1}{n}\sum_{i=1}^{n} x_i$. The method of moments estimate for μ is determined by setting μ equal to the corresponding sample moment $\overline{X}$,

$$\mu = \frac{1}{n}\sum_{i=1}^{n} x_i \tag{5.5}$$

and solving for the parameters of interest, which is trivial in this case, because μ is the parameter of interest and Equation 5.5 already presents the solution. In more complex cases, the parameters of interest may be functions of numerous moments, and multiple equations may be needed if there are multiple parameters of interest.

Note that the x_i's are specific elements of the outcome space; they do not comprise n random variables in the data model. There is one random variable X^* (to use the notation of the preceding example) that has a distribution with expected value equal to $\overline{X} = \frac{1}{n}\sum_{i=1}^{n} x_i$. Therefore, this connection alone is not sufficient to determine all information we typically require: it cannot provide information regarding variation due to the data generating process because it does not contain random variables of the data generating process. In order to obtain standard errors, we do not solely need an estimate based on the data model alone; we need an estimator. Moreover, that estimator needs to be a function of random variables associated with the data generating process, not with the data model, which being a type of population model has no sense of standard error applicable to it. Consequently, in the method of moments, we determine the estimator by substituting into the equation the random variables from the data generating process that produced each data point. So, we substitute the random variables X_i from the data generating process into the equation for each datum x_i in the data that it underlies. This yields

$$\hat{\mu} = \frac{1}{n}\sum_{i=1}^{n} X_i \tag{5.6}$$

Now Equation 5.6 is an estimator that is a function of random variables from the data generating process model, but the form of the function comes from the data model. The estimator, being a function of random variables modeling the data generating process and not data, has a corresponding standard error reflecting variation in possible estimates due to sampling.

A more complicated example of method of moments is found in determining an estimator for a linear regression function. Consider the regression function of $E(Y \mid X) = X' \cdot \beta$ defined for a data generating process model and the corresponding error term $\varepsilon = Y - X' \cdot \beta$, which has expectation equal to 0. For the data model, $\varepsilon^* = y - x' \cdot b$ in which x and y denote a matrix and vector of data values, respectively. We are interested in obtaining an estimator for the parameter β in the data generating process model. We can equate the data generating process moments of the covariance between X and ε, with its counterpart in the data model: $\text{Cov}(X, \varepsilon) = \text{Cov}(X^*, \varepsilon^*)$. Letting $\sigma_{x,\varepsilon}$ denote $\text{Cov}(X, \varepsilon)$ from the data generating process model and noting that the $\text{Cov}(X^*, \varepsilon^*)$ in the data model is $\frac{1}{n}x'(y - x \cdot b)$, we have

$$\frac{1}{n}x'(y - x \cdot b) = \sigma_{x,\varepsilon} \tag{5.7}$$

which, solving for b, gives

$$b = (x'x)^{-1}x'y - n(x'x)^{-1}\sigma_{x,\varepsilon} \tag{5.8}$$

and after plugging in the corresponding random matrix X and vector Y from the data generating process (DGP) that produced the data, we have

$$\hat{\beta} = (X'X)^{-1}X'Y - n(X'X)^{-1}\sigma_{x,\varepsilon} \tag{5.9}$$

Unfortunately, Equation 5.9 is a function of unknown parameters $\sigma_{x,\varepsilon}$. To obtain an actual estimator, we need additional information: for example if it is reasonable to presume that $\sigma_{x,\varepsilon}$ is a vector of 0s (i.e., X and ε are uncorrelated in the data generating process model) and therefore the second term on the right-hand side of the above equation is 0, we get the usual ordinary least squares linear regression estimator for β:

$$\hat{\beta} = (X'X)^{-1}X'Y \tag{5.10}$$

This may seem a bit complicated. A simpler approach is to initially consider the covariance of the random variables X with ε to be 0 in the data generating process model (what is called the *moment condition*), impose the same constraint on the data model, solve for the data model's analogous parameter (in this case b), and then substitute the DGP's random variables that generated the data into the equation to produce the usual ordinary least squares estimator.

Models of Distributions and Densities

In Chapter 4, I mentioned that an advantage of cumulative distribution functions and probability density functions is that, unlike probability measures, they are expressed as functions of the range of random variables rather than functions of sets in sigma-algebras. This advantage is particularly evident when we seek to model these functions. Mathematical models of real-valued variables are extremely common, relatively easy to work with, and likely to be familiar to most researchers.

Suppose we have a probability model of a data generating process $(\Omega, \mathcal{A}, p)$ on which we have defined a random variable X. The probability measure p implies X has some distribution F, but we may not know F (otherwise, we would not need to collect data). Moreover, although, as discussed in the preceding section, F^* from our data model will converge to F as n goes to infinity, it is not likely to be a simple function of the data's random variable X^*, and most certainly not in most realistic finite samples. Suppose we draw one million samples from a uniform distribution on the unit interval. Notwithstanding the distribution of the data generating process, the distribution of the data itself, F^*, is not likely to have a uniform distribution. Figure 5.2 shows a histogram of a variable (labeled u) from such a sample. Note that the bars of the histogram vary around 1, sometimes a little high, sometimes a little low—not by much, but they vary nonetheless. The distribution F^* of the data model is not uniform. If it were, it would be a constant, not plus or minus a little. Indeed, trying to capture F^* from Figure 5.2 may require a fairly complicated equation of u. Looking at the uneven sawtooth pattern, it would likely require an extremely high-order polynomial to capture this interval exactly—if it were possible at all (note that our density must work

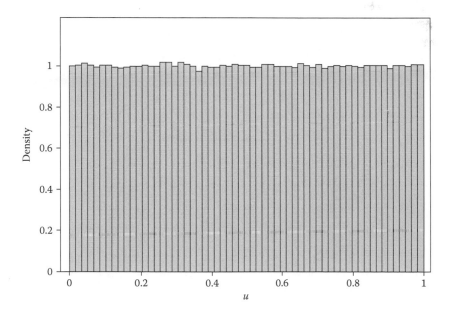

FIGURE 5.2
Histogram of one million draws from a uniform distribution on the unit interval.

point by point and not just for the arbitrary bins graphically presented in the figure). However, we may consider the variability around 1 to be sufficiently small to ignore and instead choose to model F^* as $M^*(u) = u$ (i.e., to model F^* as a uniform distribution on the unit interval). In this case, as we expect of models, our model is wrong in some respects but presumably accurate enough in some other respect that serves our purpose.

What is our purpose? Well, of course, that depends on what we are doing, but such models are commonly used to inform our understanding of F—the distribution of the random variable from the data generating process. Does this mean we presume F has a simple functional form and that since F^* converges to F, so will an appropriately specified model M^* of F^*? Although this is perhaps a common approach, it is not necessary; indeed, it would be safer to not make such a presumption.

Because a model in this case will not perfectly capture that which is being modeled, there must be criteria by which we select our model and determine whether it is sufficiently faithful in representing the features of interest. Applying a criterion to modeling F and optimizing across possible models with respect to that criterion would identify some model M reflecting features of F that we presumably seek. (I am glossing over the mathematical requirements, the technical details of which are beyond the scope of this book.) Unfortunately, as with F, we do not have M. However, if M^* is determined by the same criterion that we would apply to M, if only we could, then, as presented in Figure 5.3, we might reasonably expect M^* to converge to M as n goes to infinity. Suppose that a data generating process is represented by the probability space $(\Omega, \mathcal{A}, p)$, which implies a distribution F for our random variables of interest; using criterion c we would model F, or some relevant feature, as M_c. If we engage the data generating process n times, we obtain a data set D modeled with the probability space $(D, \mathcal{F}, P^*)$, which implies distribution F^* for its corresponding random variables. Applying criterion c, we can model F^* as M_c^*. As n goes to infinity, if F^*

$$\begin{array}{cc} \text{DGP} & \text{Data} \\ \text{model} & \text{model} \end{array}$$

$$(\Omega, \mathcal{A}, p) \Rightarrow (D, \mathcal{F}, P^*)$$

$$F \xleftarrow{\;n \to \infty\;} F_n^*$$

$$M_c \xleftarrow{\;n \to \infty\;} M_c^*$$

FIGURE 5.3
Depiction of the connection between a data generating process (DGP) and a data model. Data realizations from the data generating process are modeled by P^*. The distribution of random variables, F^*, defined on the data model converges to the distribution of the random variables that generated the data, F, as the sample size increases. Consequently, a model based on criterion c, M_c^*, of F^* converges to a similarly specified model M_c of F as the sample size increases.

converges to F, then M_c^* converges to M_c. Consequently, our modeling process applied to data will converge to the model we would achieve if applied to F and our model of data features inform features of the data generating process.

Unfortunately, in applied research we work with finite samples, and M_c^* is data specific. The closer we model F_n^*, the more likely it will *not* represent M_c but take on the detailed characteristics of F^* from the data instead of F from the data generating process—this is called *over-fitting* of data. One approach to this problem is to use a fit criterion such as the Akaike Information Criterion (AIC) or the Bayesian Information Criterion (BIC) (among many others). This approach, which I will not discuss further, is to apply a criterion that penalizes the complexity of the model to avoid selecting one that matches F_n^* too closely.

Another approach is to statistically test whether the data are inconsistent with a hypothesized F. However, to do so in practice is to actually test a model M of F (or parameter thereof), which leads to a conundrum. Because we assume, *a priori*, that M is in fact not F but is only an approximating model, the test that F is M is unnecessary: we already presume M is not F and given enough information we could confirm this. How then can we proceed to use data as evidence to investigate our questions about F? Some suggest nonparametric statistics, which avoid much of the modeling burden, but such a strategy can come at a great cost in terms of untestable assumptions, the need for larger data sets, and sometimes greater computational complexity. Although in some situations using a fit criterion or nonparametric statistics may be reasonable, I suggest we need not abandon all hope in testing parametric models, but we need to have a better understanding of what is achieved by doing so.

Rather than seek to identify F as it is in its intimate detail, we can seek to identify the models of F that are consistent with results from the data generating process (Ω^n, $\mathcal{A}^n$, p) and rule out models that are inconsistent with data. This strategy does not consider a model as a hypothesized F (we assume a model is not in fact F) but instead considers a model as a counterfactual approximation to F. Note here that we are considering the full data generating process and are thereby concerned with n as well as F. Essentially we are concerned with identifying the models of F that a given sample size cannot rule out, called *statistically adequate models*. We presume, of course, that a larger sample will garner greater refinement in discerning models, but for any given sample size the data may well be consistent or inconsistent with various models. Moreover, for smaller sample sizes the data will be consistent with more models (i.e., fewer models can be ruled out by statistical testing).

Figure 5.4 presents data from a data generating process with a sample size of 100. Table 5.5 presents tests for three nested regression models: a linear function of X (Block 1), a quadratic polynomial (Block 2), and a cubic polynomial (Block 3). Because these tests of a polynomial require comparison to the next highest polynomial, a fourth-order polynomial (Block 4) is

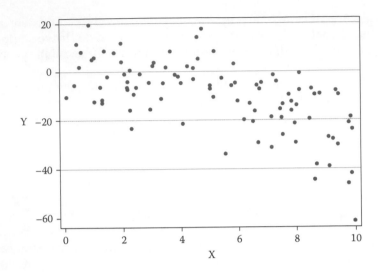

FIGURE 5.4
Scatter plot of Y versus X of 100 samples.

TABLE 5.5

Test Results for Nested Regression Models

Block	F	Block df	Residual df	Pr > F	R^2	Change in R^2
1	62.97	1	98	0.000	0.39	
2	10.75	1	97	0.002	0.45	0.06
3	0.71	1	96	0.401	0.46	0.01
4	0.03	1	95	0.875	0.46	0.00

included to facilitate testing the cubic model; also note that Block 1, being the linear model, is used to test whether the regression is merely a constant. The data provide evidence to rule out a constant mean and a model of the regression function for Y as a linear function of X, but not sufficient evidence to rule out a quadratic polynomial of X, or a cubic polynomial of X (note that I do not take insignificant findings as evidence for hypotheses being tested). In this case, we may infer that F is sufficiently different from a constant mean (the test reported for Block 1) and a linear model (the test reported for Block 2), such that a data generating process with sample size of n can discern it, but F is not sufficiently different from the higher-order polynomial models that the data generating process can discern this.

If the p-value of a specification test for M is small, then either the data are a rare case from an F that is approximately M, or the data are a common case from an F that is not approximately M. Because, by definition, it is likely that the data are a more common case than a rare case, it is reasonable to presume that F is sufficiently dissimilar to M and therefore rule out M as a statistically adequate model.

It may be that the data are consistent with a nonsingleton set of models: in the preceding example they are consistent with the higher-order polynomials. Adjudicating among such a set is typically accomplished by using criteria accounting for fit with the data and/or functional simplicity. However, because no model in this set was empirically ruled out and is thereby consistent with the data, we should be careful picking just one by nonstatistical criteria without investigating whether the substantive inferences based on these models differ from that of the model we otherwise prefer.

This logic of inference extends to testing functions of parameters associated with a statistically adequate model. Suppose we identify M as a statistically adequate model and seek to use it for testing a hypothesis regarding F. If the p-value is small, based on M, then F must be sufficiently different than the hypothesized characteristic to warrant ruling out that hypothesis. For example, suppose we have the following hypothesis regarding F:

$$\frac{\partial E_F(Y|X)}{\partial X} > 0 \qquad (5.11)$$

If M is a statistically adequate model and a test of whether

$$\frac{\partial E_M(Y|X)}{\partial X} > 0 \qquad (5.12)$$

garners a small p-value, then we have evidence that, although M is a statistically adequate model of F, the characteristic $\partial E_F(Y|X)/\partial X$ is sufficiently different from the hypothesis that the data from a data generating process with sample size n can discern this fact. We can then rule out the hypothesis as characterizing F even though it was based on a test of M. If the p-value is large, then the characteristic is not sufficiently different from the hypothesis that it can be discerned by the data generating process, and therefore it cannot be ruled out. The statistical test is based on the counterfactual M rather than F: we ask how likely it is that we would get data at least as extreme as what we observed if F was M, even though we presume it is not. However, we use this information to draw inferences regarding F.

It is important to note that according to this logic, data provide stronger evidence by virtue of ruling out models or characteristics of a distribution. Data provide disconfirming evidence for a model or characteristic by ruling it out, and data provide confirming evidence for a model or characteristic by not ruling it out while ruling out the alternatives. Data do not provide evidence to adjudicate among models that are all consistent with the data. Consequently, disconfirming evidence is typically stronger than confirming evidence. If many models or characteristics are ruled out, then each model, individually, is taken to be unlikely. However, if many models or characteristics are not ruled out (i.e., they are statistically adequate), then each model is only one of many candidates. Fit criteria may be used to distinguish these models, but their epistemic value does not accrue from an understanding of the underlying data generating process.

Arbitrary Models

The preceding models provide (simplified) examples of probability models in common usage. However, because mathematical probability is a calculus awaiting interpretation, we can apply it in other ways that may be useful. Suppose I wish to investigate characteristics of the United States House of Representatives. I might proceed with a population model defined with an outcome set Ω comprising the members of the House, the power set of Ω as the event set $\mathcal{A}$, and a probability P that assigns $1/N$ to each unit set (each individual member) in which N denotes the number of members of the House. In this case, $(\Omega, \mathcal{A}, P)$ would be a probability space that would allow me to characterize the normalized histogram of random variables measured on members of the House—that is, the population model of the House of Representatives.

If we define two random variables on the corresponding measurable space as X denoting a measurement of each member's support for universal health care and Y denoting the measurement of each member's years in Congress divided by the total person-years across all members of Congress (perhaps taken to be a measure of relative influence in the House). Then the expectations $E_P(X)$, $E_P(Y)$, and $E_P(X \cdot Y)$ represent the average support for universal health care across the House, the average influence across the House, and the average effective support for universal health care in the House.

Suppose, however, that I am instead interested in investigating total support for universal health care. I may then use Y (defined as a measure on $\mathcal{A}$) as my probability rather than $1/N$ and define my probability space as $(\Omega, \mathcal{A}, Y)$. Now $E_Y(X)$ becomes the total influence-weighted support for universal health care. Consider the difference:

$$E_P(X \cdot Y) = \frac{1}{N} \sum_{\Omega} X_w \cdot Y_w \tag{5.13}$$

whereas

$$E_y(X) = \sum_{\Omega} X_w \cdot Y_w \tag{5.14}$$

The expectation associated with the population model in Equation 5.13 provides the average effective support for universal health care among the members of the House, whereas the expectation in Equation 5.14 associated with $(\Omega, \mathcal{A}, Y)$ provides the effective support for universal health care.

Of course, in this case we can use $(\Omega, \mathcal{A}, P)$ to get the same result by taking the expectation of a variable $W = N \cdot X \cdot Y$, but it is less intuitive and ad hoc.

By using $(\Omega, \mathcal{A}, Y)$, it is easy to evaluate effective support directly. Suppose X has a range of $[-1, 1]$, in which -1 corresponds to full support for efforts against universal health care, 1 corresponds to full support for efforts toward universal health care, and 0 corresponds to no support either way. The effective support for efforts in favor of universal health care is the probability (again, defined in terms of Y) of X greater than 0, whereas the effective support for efforts against universal health care is the probability of X less than 0.

Suppose we define another random variable Z as a measurement of support for efforts to decrease federal power. Analysis based on our population model $(\Omega, \mathcal{A}, P)$ may result in a decreasing regression function of $E_P(X \mid Z)$, as shown in Figure 5.5a, indicating that the average support for universal health care decreases with support for decreasing federal power. However, an analysis based on $(\Omega, \mathcal{A}, Y)$ could result in an increasing regression function of $E_Y(X \mid Z)$ as shown in Figure 5.5b, in which the size of the dots reflects the relative influence of the House member. By this analysis, the effective support for universal health care is positively related to support for decreased federal power. This difference is driven by the correlation between X and Z among those who have more years of experience in the House.

Another example is the use of probability spaces to help conceptualize the explanatory power of a theory. Suppose we wish to explicitly define

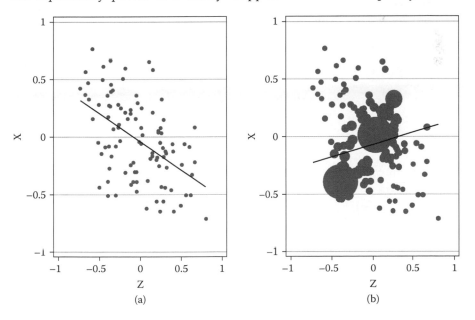

FIGURE 5.5
(a) The relationship between X and Z based on $(\Omega, \mathcal{A}, P)$; (b) the relationship between X and Z based on $(\Omega, \mathcal{A}, Y)$. The solid lines in each panel represent the regression lines.

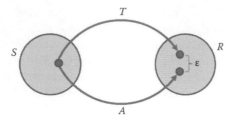

FIGURE 5.6
Two maps, T and A, from S to R in which ε represents a difference, assuming R is a metric space.

explanatory power. We could do so in different ways but each drawing on the same initial representation. Define S to be the set of real-world situations that fall within the scope of a given theory. Also define R to be another set of situations that will represent possible results that follow from each situation in S. We may then consider a map $T: S \rightarrow R$ representing the resulting situation predicted by the theory for each situation in S. We may also define a map $A: S \rightarrow R$ representing the actual resulting situation that follows from each situation in S (see Figure 5.6).

To clarify our understanding, we can define a probability space $(S, \mathcal{A}, P)$ for which $\mathcal{A}$ is the power set of S and $P(\{s\}) = 1/|S|$. We can define a second measurable space $(R, \mathcal{F})$ and two random variables T and A measurable $\mathcal{A}$. Consequently, the vector (T, A) is a random vector from $(S, \mathcal{A}, P)$ to $(R^2, \mathcal{F}^2, p)$ in which p is the corresponding probability defined in terms of P. Based on this representation, we can provide definitions for explanatory power in terms of agreement (the extent to which T and A agree on the results for situations in S) or disagreement (the extent to which T and A disagree regarding the results for situations in S).

Regarding agreement, we might define

$$\text{Power}_1 = \sum_{r \in R} p(T = r, A = r) = \sum_{r \in R} P(T^{-1}(r) \cap A^{-1}(r)) \qquad (5.15)$$

Power$_1$ sums the intersections of sets $T^{-1}(r)$ and $A^{-1}(r)$ in S across all r in R. Figure 5.7 shows these sets for an arbitrary r. The definition of power in Equation 5.15 provides the proportion of situations in S for which the theory agrees with actual results.

Alternatively, we can create a definition based on disagreement by defining another random variable $D = (T(s) - A(s))^2$ measurable $\mathcal{F}^2$ on $(R^2, \mathcal{F}^2, p)$ to $(\mathbb{R}, \mathcal{B}, p_D)$ in which $\mathbb{R}$ is the real line, $\mathcal{B}$ is the Borel sigma-algebra, and p_D is derived from the preceding probability spaces. We can then define

$$\text{Power}_2 = E_{p_D}(D) \qquad (5.16)$$

Power$_2$ represents the expected squared difference in results between what the theory predicts and what would actually happen.

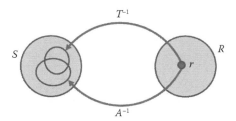

FIGURE 5.7
The overlapping subsets in S corresponding to the pre-images of T and A corresponding to a single point r in their range.

We can see that Power$_1$ is simpler to understand than Power$_2$ because it directly represents agreement, which matches the colloquial sense of explanatory power (an increase in Power$_1$ corresponds to what we would consider an increase in explanatory power). On the other hand, Power$_2$ is interpreted in the opposite direction (an increase in Power$_2$ is interpreted as a decrease in explanatory power). This analysis provides another reason, however, for favoring Power$_1$ over Power$_2$, one that does not turn on semantics (we could have sought a concept for explanatory deficit): Power$_2$ requires a more complicated random variable D that allows for differences to be calculated (which can be thought of as requiring a distance metric defined on R rather than merely allowing it to be a set). Power$_1$ does not require us to define such a distance metric.

We could have defined the indicator of whether T and A map to the same result as $D = \mathbf{1}(T(s) = A(s))$. The measurable space that D maps into could then be simplified to $(\{0, 1\}, \mathcal{P}, p_D)$, where $\mathcal{P}$ is the power set of $\{0, 1\}$: that is, $\{\{0, 1\}, \{0\}, \{1\}, \varnothing\}$. This definition of D avoids the requirement of a distance metric and makes the interpretation of Power$_2$ the same as that of Power$_1$: the proportion of situations in S for which T and A agree.

The main point here is not that such models are in common use but rather to emphasize that, from the mathematical perspective, probability spaces are uninterpreted objects that carry with them the full scope of probability results. The utility and meaning of those results depends solely on the interpretation of the probability space that the analyst provides—in other words, meaning depends on what the probability space is being used to model and is not inherent in the mathematical definition of probability itself.

Additional Readings

This chapter provided examples of how probability spaces can be used to model some specific research problems based on my perspective that probability is a mathematical structure that the researcher can use under any useful interpretation. I answered the question of what probability is by reference to its mathematical definition. What it can be used for is only limited by the imagination of the user, who need only provide a coherent interpretation and legitimate rationale for its application. There is, however, considerable literature on differing views regarding the proper answer to the question regarding what probability is. These views take a substantive meaning as definitional of the term and consider mathematics as constraining how it is applied. Consequently, if there is to be only one substantive meaning of probability, it should not be a surprise that there is considerable argument over what that definition ought to be.

Gillies' book *Philosophical Theories of Probability* (Routledge, 2000) and Mellor's *Probability: A Philosophical Introduction* (Routledge, 2005) both provide fairly easy-to-read overviews to the primary competing theories. The books *Probability is the Very Guide of Life: The Philosophical Uses of Chance*, edited by Kyburg and Thalos (Open Court, 2003), and *Philosophy of Probability: Contemporary Readings*, edited by Eagle (Routledge, 2011), provide compilations of papers discussing some key contested issues.

There are also books that consider the various definitions of probability in the context of science, such as Suppes' book, titled *Representation and Invariance of Scientific Structure* (CSLI Publications, 2002). Moreover, there are books that consider, and advocate, various definitions of probability for the purpose of accruing and using scientific evidence. Earman's book, titled *Bayes or Bust? A Critical Examination of Bayesian Confirmation Theory* (MIT Press, 1992), gives an analysis of Bayesian theory as an epistemological tool for accruing evidence. Mayo's book, titled *Error and the Growth of Experimental Knowledge* (University of Chicago Press, 1996), provides a defense of the frequentist definition of probability. Similarly, *Error and Inference: Recent Exchanges on Experimental Reasoning, Reliability, and the Objectivity and Rationality of Science* (Cambridge University Press, 2010), edited by Mayo and Spanos, provides an updated defense of the frequentist definition of probability and the consequent statistical tools used in science. This book is in the form of position chapters by Mayo and Spanos, challenges from various contributing authors, and responses by Mayo and Spanos. The book *The Nature of Scientific Evidence: Statistical, Philosophical, and Empirical Considerations* (University of Chicago Press, 2004), edited by Taper and Lele, provides a compilation of chapters that argue for specific approaches to scientific evidence—many explicitly adopting a specific definition of probability, while others are less explicit in doing so.

For an excellent book on method of moments and its generalization, see Hall's *Generalized Method of Moments* (Oxford, 2005).

6

Common Problems

In this chapter, I will address some vexing problems that motivated writing this book. The structure of this chapter is different from the preceding ones in that much of the presentation is framed by presenting problem statements and providing corresponding solutions in which I clarify the issues and the approach to resolving them.

This chapter addresses some of the most challenging conceptual errors that applied researchers face. You will likely see many of them as extensions of the topics discussed in the preceding chapter. However, here they will be elaborated upon within the context of a research problem.

Interpreting Standard Errors

By now it should be clear that in statistics, all probability-related characteristics of random variables stem from the underlying probability space; consequently, the interpretation of such characteristics comes from the interpretation of the underlying probability space. If you are using a probability space to model uncertainty of an objective process by which the data are generated, then the standard deviation of your random variables represents variations associated with the possible realizations of the random variable from that process. In other words, from a strictly frequentist theory of probability you would interpret your random variables in terms of the values you could get if the process were repeated an infinite number of times; from a propensity theory of probability, you would interpret your random variables in terms of the inherent potential of a given process to produce particular values. If you used a probability space to model your subjective uncertainty about some outcome, answer, or quantity, then you would interpret your random variables in terms of your beliefs about the potential value of the quantities being considered.

A statistic, being a function of random variables, is itself a random variable; therefore, it has a distribution associated with its underlying probability space. The standard error of a statistic is typically defined as its standard deviation (assuming it has one), our estimator of which is often used in conjunction with the original statistic to define another statistic with a known distribution, or known limiting distribution of a function of the statistic.

This definition is suitable for most purposes, but we run into some difficulty in more complicated situations, particularly when the underlying probability measure can be factored into a part that represents a data generating process and a part that does not. This will become clear as we discuss the problems presented below.

To frame this distinction, consider that the phrase "standard error" is meaningful solely in terms of capturing a consequence of a data generating process. I interpret the standard error as reflecting the variation in a statistic due to the possibility of obtaining different data sets from a data generating process. For example, suppose I sample patients associated with each of the primary care physicians of a given healthcare organization. Physicians are not sampled because all physicians are determined to be used; only patients are sampled. My probability space can be defined as $(\Omega \times S, \mathcal{A}_\Omega \otimes \mathcal{A}_S, P)$ in which Ω denotes the population of patients and S denotes the set of physicians in the organization. The probability measure for an arbitrary patient–physician pair (w, s) can be factored as $P(\{(w, s)\}) = P(\{w\} \times S \mid \Omega \times \{s\}) \cdot P(\Omega \times \{s\})$ in which the conditional probability reflects the probability of the data generating process obtaining individual w from a physician s, and the marginal probability is the probability associated with the population model of physicians. In other words, setting $P(\Omega \times \{s\}) = 1/N_S$ reflects the set of N_S physicians and allows for representing their normalized histogram of characteristics. It does not model a data generating process and is not, therefore, a sampling probability. Consequently, sample variation of any random variable is due solely to the part of the probability measure, that is, modeling the data generating process (i.e., due to $P(\{w\} \times S \mid \Omega \times \{s\})$), and not due to the part modeling the physician population (i.e., not due to $P(\Omega \times \{s\})$).

The preceding example shows that the standard deviation of an estimator may not be its standard error defined to reflect variation due to a data generating process. The standard deviation will reflect the variance derived from the overall probability measure, P, whereas the standard error should only reflect the source of variance due to the possible outcomes of generating the data, $P(\{w\} \times S \mid \Omega \times \{s\})$ in this example. The standard error should be interpreted as reflecting how much we roughly expect an estimate to deviate from the estimator's mean value *due to the variation in possible samples we could obtain from a given data generating process.* If a standard error of an unbiased estimator for mean blood pressure is 5, then we roughly expect the estimate to be wrong by approximately 5 units due to sampling. I use the phrase "roughly expect" because the standard error is based on the square root of an expected value and not the expected absolute deviation directly.

To drive this point home, the standard deviation is a mathematical property of a probability distribution, and it retains its meaning regardless of whether the probability space is modeling something of substantive interest to a researcher or modeling nothing substantive as might interest a pure mathematician. The standard error, on the other hand, is strictly a contextually interpreted concept; it has no meaning in the absence of its representation of variation due to a data generating process. The standard deviation and the standard error are the same only in the case that the underlying probability measure is solely representing variation in the data generating process, which is the most common case in applied research. Consequently, one cannot properly identify a standard error without first knowing what the probability measure of a probability space is modeling.

In summary, if you have not thought about what you are modeling with the probability space that underlies your random variables, then they are not interpreted, and it is hard to tell what makes your variables random and what the appropriate standard errors are. You may then make the mistake that some do when faced with the mistaken argument in the following problem.

Problem 6.1

Critique the following passage from Goodman SN (1999) "Toward Evidence-Based Medical Statistics. 1: The P Value Fallacy" *Ann Intern Med.* 130:995–1004.

"A classic statistical puzzle … involves two treatments, A and B, whose effects are contrasted in each of six patients. Treatment A is better in the first five patients and treatment B is superior in the sixth patient. Adopting Royall's formulation (6), let us imagine that this experiment was conducted by two investigators, each of whom, unbeknownst to the other, had a different plan for the experiment. An investigator who originally planned to study six patients would calculate a p-value of 0.11, whereas one who planned to stop as soon as treatment B was preferred (up to a maximum of six patients) would calculate a p-value of 0.03 (Appendix). We have the same patients, the same treatments, and the same outcomes but two very different p-values (which might produce different conclusions), which differ only because the experimenters have different mental pictures of what the results could be if the experiment were repeated. A confidence interval would show this same behavior."

Solution: Goodman makes the fallacy of misclassification in his presumed argument. His argument is based on a model of subjective uncertainty, whereas the two investigators presumably based their decisions on models of data generating processes. The standard errors associated with the first investigator's estimator reflect the variation in the effect measure due to the potential for a given process that samples six subjects. The standard errors associated with the second investigator's estimator reflect the variation in the effect measure due to the potential for a given process that samples subjects until treatment B is preferred. These are indeed distinct processes that in fact have different standard errors. Goodman considers the difference to be in the mental pictures of the investigators, from which I infer Goodman is thinking of a subjective measure. While Goodman contends the results "differ only because the experimenters have different mental pictures of what the results could be if the experiment were repeated," I imagine the investigators would contend that the results differ because they were using different procedures, and each procedure in fact has a different distribution of possible outcomes. If you doubt this, set up a Monte Carlo experiment in which you could simulate each of these two processes. You will find the standard errors are in fact different—regardless of your mental picture. Goodman's concern over having "the same patients, the same treatments, and the same outcomes but two very different p-values" exhibits confusion between data, which are not random variables in the context of the investigation, and the actual random variables that reflect the data generating process. Moreover, there is a lack of understanding that data cannot change the characteristics of random variables representing a data generating process.

Notational Conventions

Before addressing key issues that measure theory can help us sort out, we need notational conventions. For the purpose of the remainder of this book, it will help to simplify our notation for probability measures defined on measurable spaces having product sets for their outcomes and to have a notational language that clearly and coherently uses indexing of variables and values.

In Chapter 4, conditional probabilities on a probability space such as $(\Omega \times S, \mathcal{A}, P)$ are denoted as $P(\{w\} \times S \mid \Omega \times \{s\})$ to indicate the probability of obtaining an outcome containing w given the outcome contains s, and the marginal probability of an outcome containing w can be denoted as $P(\{w\} \times S)$. For greater concision, however, in the rest of this chapter these will be denoted as $P(\{w\} \mid \{s\})$ and $P(\{w\})$, respectively. Although technically incorrect, this notation will make it easier to more concisely represent the problems addressed.

A variable is a function that is denoted by capital letters. Sets are also denoted by capital letters. The context of their use will distinguish them. For example, a function X from some domain D to some range R is a variable representing the mapping $X: D \rightarrow R$. A vector or matrix of variables will be denoted as a bold capital letter, such as $\mathbf{X}$.

The value of a variable applied to an arbitrary element of its domain is denoted by a lowercase letter with a subscript indicating the arbitrary element. For variable X with domain D, the value of X applied to element $d \in D$ is $x_d = X(d)$. In this case, x_d is the specific value in the range of the variable X evaluated at the element d in D. The equal sign appropriately denotes the mathematical equivalence of the left- and right-hand sides of the preceding equations. Nonetheless, the left- and right-hand sides are conceptually different, just as $X(d) \equiv d^2$ leads to $4 = X(2)$ in which the left-hand side is a number and the right-hand side is a function evaluated at a number. The meaning of the squared function evaluated at 2 is not the same as the meaning of the number 4.

We often choose different letters or words to denote different variables; for example, three variables may be denoted by X, Y, and Z. Alternatively, particularly when we wish to denote many variables, we can differentiate variables by an index rather than letters. Instead of a set of three variables being denoted as $\{X, Y, Z\}$, we may denote them as $\{X_1, X_2, X_3\}$. More generally, we may specify an index set and specify individual variables by virtue of the indices rather than different uppercase letters. For example, suppose we determine an index set of $I = \{1, 2, 3, 4, 5, 6, 7, 8, 9, 10\}$, the set of variables $\{X_i : i \in I\}$ contains 10 separate variables. An index set need not be numerical, and if it is numerical it need not be either discrete or finite. However, for our purpose, they will usually be numerical and both discrete and finite.

To be useful in applied research, the index set is often defined as the range of a one-to-one function that maps from a conceptually understood domain. The inverse of the function thereby provides an interpretation of the index set. For example, suppose I wish to index variables associated with each member of a set Ω that comprises individuals of a particular population. Then, a one-to-one index function $\mathcal{J}^{-1}: \Omega \to I$ defines I as an index set for Ω, and $\mathcal{J}: I \to \Omega$ provides an interpretation for each element $i \in I$ in terms of the meaning ascribed to Ω. As another example, suppose I define O as the set of data generating processes and I_N the set of numbers from 1 to N, where N is the number of elements in O. The function $\mathcal{J}^{-1}: \Omega \to I_N$ defines I_N as the index set for N data generating processes, that is, the set of occurrences of the data generating process that will ultimately produce a data set of N results. $\mathcal{J}: I_N \to \Omega$ provides an interpretation for each element $i \in I_N$ as the observation number associated with a particular data generating process.

Note that for clarity I write the interpretation function as $\mathcal{J}$ and the indexing function as its inverse $\mathcal{J}^{-1}$ rather than the reverse as might be expected given their introduction above. This is because it will be easier to speak of the interpretation function and therefore I use the simpler notation to denote it. Being a one-to-one function, this is arbitrary. Suppose Ω is a set denoting three people {Fred, Ethel, Hildegard} and a corresponding index set is $I = \{1, 2, 3\}$. An interpretation of I could be $\mathcal{J}(1) = $ Fred, $\mathcal{J}(2) = $ Ethel, and $\mathcal{J}(3) = $ Hildegard. In this case, a variable X_2 would be interpreted as the variable X associated with Ethel.

Note that $\mathcal{J}: \Omega \to \Omega$ is possible and can be useful when arbitrary, rather than explicit, indices are being discussed. However, it can be useful to index a nonnumerical set by a numerical one; for example, if Ω is nonnumerical (e.g., a set of people who compose a population of interest), then it can be useful to index it with a numerical set I (e.g., $\mathcal{J}^{-1}: \Omega \to I$) thereby having an interpretation $\mathcal{J}: I \to \Omega$, as in the example presented in the preceding paragraph.

The advantage for the applied researcher of using the same uppercase letter (or letters, or word, or even words) in denoting different variables via an index i is not merely efficiency in denoting many variables, but also in allowing the researcher to use the uppercase component to denote a specific concept and leaving it to the indices to denote the different variables. For example, suppose I wanted to define variables that captured blood pressure and stress level for 10 different people. I may use the letter B to denote the concept of blood pressure and S to denote the concept of stress. Defining a set I that indexes the 10 people, I can denote variables for blood pressure and stress associated with each person as B_i and S_i for $i \in I$. Therefore, the symbols for each variable have conceptual information that can carry throughout an analysis. In this example, wherever we see a "B" we know it is referring to a random variable representing blood pressure, and wherever we see an "S," we know it is referring to a random variable representing stress.

Variables can also be distinguished with multiple indices, as long as the combination of indices provides a unique (one-to-one) identification: $\mathcal{J}^{-1}$: $\Omega \rightarrow I \times T$ with corresponding interpretation $\mathcal{J}$: $I \times T \rightarrow \Omega$. For example, we can use two sets $I = \{1, 2, \ldots, N\}$ and $T = \{1, 2, \ldots, T\}$ to identify a set of variables $\{X_{it} : i \in I, t \in T\}$. Suppose I wish to define variables that represent the temperature at each point of the earth's surface. I can define a set O that indexes longitude and a set A that indexes latitude. Note that in this case both index sets are continuous. With these indices, I can then define a set of variables $\{T_{oa} : o \in O, a \in A\}$ that represent these temperature variables. As another example, the indexing may be hierarchically structured as shown in Table 6.1. Rather than a single index i with index set I, two indices j and k, with index sets J and K, are used such that k is nested within j. Note in the table that each element of Ω is uniquely indexed by the (j, k) pair.

If, unlike in Table 6.1, the number of k elements associated with each i index varies, then we would require a set of index sets, one for each i: $\{K_i : i \in I\}$.

By this indexing scheme, if we were to denote a specific variable such as for $i = 2$ and $j = 3$, we would have T_{23}, which could mistakenly be interpreted as an index value 23 for a single indexed variable T_i. We require additional notation to avoid this ambiguity, such as putting a comma between the numbers; however, in this text, in which we will work only with arbitrary indices

TABLE 6.1

Single and Paired Index Sets

Ω	I	J, K	
	i	j	k
w	1	1	1
w'	2	1	2
w''	3	1	3
w'''	4	2	1
w''''	5	2	2
w'''''	6	2	3

denoted by single letters, we can dispense with additional notation for this purpose.

For random variables defined on probability spaces, we will add one more notational nuance. Remember that when a data generating process underlies our collection of N samples, we represent it by a probability space such as $(\Omega^N, \mathcal{A}^N, P)$ that comprises the N probability spaces $(\Omega, \mathcal{A}, P_i)$ for each $i \in \mathbb{I}$ in which the interpretation of $\mathbb{I}$ is the set of processes that will each produce an outcome from Ω. Each of these component probability spaces will have random variables defined on them that we may wish to distinguish via indexing by i. But suppose that we wish to use indexing to further distinguish the variables within the observation. For example, we may be measuring blood pressure at three different fixed times on any person w that the data generating process obtains. We could denote them as $Y1_i$, $Y2_i$, and $Y3_i$ or we could use an index set $T = \{1, 2, 3\}$ interpreted as the set of times and denote them as Y_{it} for $t \in T$. However, it will become useful to distinguish the observation indexing (using i in this example) from the within-observation indexing (using t in this example). We do this by separating them with a comma: for example, $Y_{i,t}$. In a more general sense, suppose we use two indices (e.g., i and k) to identify observations (i.e., individual runs of a data generating process) and one index to identify repeated measures (e.g., t), our random variables would be written as $Y_{ik,t}$. The indexing to the left of the comma will typically reflect the structure of the data generating process, whereas the indexing to the right of the comma will distinguish variables within observation.

This nuanced notation provides a different meaning for Y_{it} and $Y_{i,t}$. The first can represent an observation defined by a random selection of a person and then a random selection of a time at which to measure Y for that person. Note that t is not specifying the time in this case but an index of the occurrence, or operation, of selecting a time. The second can represent a random selection of a person and the measure of Y at a predetermined time.

So far, we have identified two types of indices, one that applies to variables, which denotes distinct variables, and one that applies to values, which denotes arbitrary elements of a variable's domain. We can add nuance if needed. Consider two variables X_1 and X_2 with the same domain:

$$X_1: D \to R \tag{6.1}$$

$$X_2: D \to R \tag{6.2}$$

Using the preceding, indexing for arbitrary values of the variables in Equations 6.1 and 6.2 leads to the notation

$$x_d = X_1(d) \tag{6.3}$$

$$x_d = X_2(d) \tag{6.4}$$

which is ambiguous in the denoted values. Consequently, when using indexed variables, their corresponding values can be uniquely specified by the following notation:

$$x_{1:d} = X_1(d) \tag{6.5}$$

$$x_{2:d} = X_2(d) \tag{6.6}$$

or, for all i's in index set $\mathbb{I}$:

$$x_{i:d} = X_i(d) \tag{6.7}$$

Functions may take multiple arguments, and consequently so can variables. A variable may therefore be defined with a product set as its domain. For example, consider the function $X: D \times F \to R$. In this case, X is a variable that identifies each pair $(d, f) \in D \times F$ with an element in its range R. The corresponding value is then denoted as $x_{df} = X(d, f)$. If on the other hand, $X(d, f) = X(d, f')$ for all $f \in F$ and all $f' \in F$, then we denote $X(d, f)$ simply as $X(d)$ in contexts where this notational simplicity is clear.

We can now express notation more generally. Using all of the nuances expressed above, we would have

$$x_{ij,st:df} = X_{ij,st}(d, f) \tag{6.8}$$

which would indicate an arbitrary value and corresponding random variable with a Cartesian product having elements (d, f) in a domain and indexed by i and j to distinguish variables across observations and s and t to distinguish variables within observation. This can seem complicated, but we will usually leave off the variable index on the value (corresponding to the lowercase labels) when the context makes it clear to which variable the value applies: for example, in the preceding case, we would typically denote the value as x_{df}. It is, however, important to preserve the notational nuance for the variables (corresponding to the uppercase labels).

Variables may be functions of other variables. In this case, the arguments for the left-hand side variable must be the same as the full set of arguments on the right-hand side (except for right-hand side arguments that are integrated out). For example, consider the following equation of variables in which α and β are constants:

$$Y(d, f) = \alpha + \beta \cdot X(d) + \mathcal{E}(f) \quad \text{for all } d \in D \text{ and } f \in F \tag{6.9}$$

Although X and $\mathcal{E}$ have different domains in this example, Y has a Cartesian product for a domain (i.e., $D \times F$) that comprises the domains of the right-hand side variables. This should be intuitively clear because the variation in Y is due to the variations in both X and $\mathcal{E}$, which are due to the arbitrary elements across their respective domains. Alternatively, and more accurately, X and $\mathcal{E}$ are each function of $D \times F$, but each only varies according to one of its arguments.

Because the left-hand side domain is determined by the domains of the right-hand side variables, we can drop the left-hand side denotation of domain elements, unless needed for clarity. Equation 6.9 could simply be written as

$$Y = \alpha + \beta \cdot X(d) + \mathcal{E}(f) \quad \text{for all } d \in D \text{ and } f \in F \tag{6.10}$$

Suppose we define an index set $\mathbb{W}$, with an interpretation $\mathcal{J}: \mathbb{W} \to \Omega$ for some population Ω, and we specify

$$Y_w = \alpha_w + \beta \cdot X_w(d) + \mathcal{E}_w(f) \quad \text{for all } w \in \mathbb{W} \tag{6.11}$$

Then for each w in $\mathbb{W}$, and thereby for each corresponding element $\mathcal{J}(w)$ of Ω, there are three variables $Y_w, X_w, \mathcal{E}_w$ that are related according to the preceding equation. As we will use such equations here, we will typically consider the distinct letters Y, X, and $\mathcal{E}$ to denote distinct concepts that remain constant across the indices. For example, Y may denote out-of-pocket healthcare expenditures, X may denote income, and $\mathcal{E}$ may denote a combined influence of all other factors, or simply the difference between Y_w and the model $\alpha_w + \beta \cdot X_w(d)$. Even though when indexed by w they denote different functions for each $w \in \mathbb{W}$, thereby interpreted as different function for each $\mathcal{J}(w) = \omega \in \Omega$, they will denote the same conceptual definitions. In this example, in which the index set $\mathbb{W}$ represents a set of people, the preceding equation indicates that for each person in Ω his out-of-pocket expenditures will vary according to the level of income, which, being a variable, is itself allowed to vary.

Note that in the preceding equation the term α_w is a constant, yet it is indexed by w. This means that each w in $\mathbb{W}$ has its own constant; therefore, α_w can vary across $\mathbb{W}$, but it does not vary within w. Being an indexed constant, α_w is a value of a function that has $\mathbb{W}$ as its domain: that is, there is some function $A: \mathbb{W} \to R$ such that $\alpha_w = A(w)$, which according to the interpretation function yields $\alpha_w = A(w) = A(\mathcal{J}^{-1}(\omega)) = A^*(\omega)$, a function with domain Ω as well. Expressing an indexed constant as a function evaluated at the interpretation of the index will be important in translating structural models of variables into probability models of random variables.

Another useful notational maneuver is to apply Occam's Razor to our indices and only use the minimal number of variables required. In the present context, this means that if

$$X_{it} = X_{it'} \quad \text{for all } t \text{ and } t' \text{ in } \mathbb{T} \tag{6.12}$$

then we can create a variable $X_i = X_{it^\circ}$ for any arbitrary $t^\circ \in \mathbb{T}$ and replace each X_{it} for all $t \in \mathbb{T}$ by X_i. Procedurally, this simply means deleting any index across which the variables are the same.

When the variable domains are clear or are not the subject of discussion, we may drop their explicit denotation; for example, we may write $X_i(d)$ simply as X_i, and the preceding equation as

$$Y_i = \alpha_i + \beta \cdot X_i + \mathcal{E}_i \tag{6.13}$$

with the understanding that Y_i, X_i, and $\mathcal{E}_i$ are variables, each having an appropriate domain.

Finally, careful consideration of the preceding discussion will reveal that the following equivalencies hold for appropriate definitions of index sets and domains:

$$Y_i(t) = Y(i,t) = Y_t(i) = Y_{it}(d) \quad \text{for all } d \in D \tag{6.14}$$

with D any domain. For the last term to be equivalent means that the value of Y_{it} at each element of a domain d is constant across all $d \in D$. Consequently, the strategy for indexing is not part of the logic of mathematics, but only a way of structuring symbols so that they are useful to your purpose. What constitutes a domain or an index set is arbitrary and must be selected for a specific representational purpose.

Suppose we are interested in a person's response to different levels of an intervention (e.g., drug dosages, tax rates, or voucher levels). We define D as the set of possible intervention levels, and for each person population Ω, with index set W, at time t in an index set for time $\mathbb{T}$, and for simplicity defining Ω as its own index set, we define a variable $X_{wt}: D \rightarrow \mathbb{R}$ to be the real-valued function of the intervention level and $Y_{wt}: D \rightarrow \mathbb{R}$ to be the real-valued function of response. We consider the relationship between these variables to be

$$Y_{wt} = \alpha_w + \beta \cdot X_{wt}(d) + \varepsilon_{wt} \tag{6.15}$$

Both Y_{wt} and X_{wt} are variables for each person w and time t that can take on different values depending on the arbitrary intervention level d. This is a structural specification: it indicates that X can potentially, or hypothetically, vary within w at any time t. Clearly, it would not be an essential characteristic of w such as sex, birthdate, or age; this is because for any given w and t such characteristics could not vary. The constant α_w is a fixed value for each w and may be defined in many different ways; for this example, we could define α_w as

$$\alpha_w = \frac{1}{T} \sum_{t=1}^{T} Y_{wt}(d^0) \tag{6.16}$$

in which d^0 denotes the element of D such that $X_{wt}(d^0) = 0$, that is identified with no intervention (e.g., no drug, no tax, or no voucher). The constant α_w in Equation 6.16 is then the average across time of individual w's responses without intervention. The constant ε_{wt} can be thought of as a deviation at d^0 for all w and t

$$\varepsilon_{wt} = Y_{wt}(d^0) - \alpha_w \tag{6.17}$$

Alternatively, we could define α_w as the value of Y in the condition of no intervention at time $t = 0$,

$$\alpha_w = Y_{w0}(d^0) \tag{6.18}$$

leading to a deviation ε_{wt} of

$$\varepsilon_{wt} = Y_{wt}(d^0) - Y_{w0}(d^0) \tag{6.19}$$

for which $\varepsilon_{w0} = 0$ for all w's.

Using the structural model in Equation 6.15 with the first definition of α_w, let us consider how to apply the notational conventions to the corresponding random variables for two data generating processes.

Example 6.1

Consider a probability space $(\Omega^N, \mathcal{A}^N, P)$ in which P models a data generating process of independently selecting N observations. Consequently, each observation indexed by $i \in \mathbb{I} = \{1, 3, \ldots, N\}$ has its own probability space $(\Omega, \mathcal{A}, P_i)$, and for any pair (i, j) with $i \in \mathbb{I}$ and $j \in \mathbb{I}$ such that $i \neq j$, P_i is independent of P_j and $P_i = P_j$ for all $\{w\} \in \mathcal{A}$. Note that the interpretation of $\mathbb{I}$ is the set of N occurrences of a data generating process with independent observations.

For simplicity of this initial example, let us consider t to be fixed for all w—this is a repeated measures observational design in which we measure our variables at a fixed set of times (perhaps a baseline and a six-month measure). Define $d_{i,t:w} = D_{i,t}(w)$ as the intervention level for each w measured at the time t. Note that our random variables are defined on $(\Omega, \mathcal{A}, P_i)$ and are thereby functions of the domain Ω and not of D, the domain of the structural model. The following definitions of random variables correspond to our structural variables:

$$Y_{i,t}(w) = Y_{wt}(d_{i,t:w}) \tag{6.20}$$

$$X_{i,t}(w) = X_{wt}(d_{i,t:w}) \tag{6.21}$$

These produce values of Y and X associated with the value d_w that is present at the time of observation. Note that the indices of the random variables are separated by a comma—this indicates that the i index refers to different observations (as determined by the data generating process), whereas the t index refers to different within-observation variables, which in this case are different preset times.

Note that our constants in the structural model vary across indices, which can be expressed as values of variables defined on the corresponding index sets:

$$\alpha_w = A_{i,t}(w) \tag{6.22}$$

$$\varepsilon_w = \mathcal{E}_{i,t}(w) \tag{6.23}$$

However, α_w being constant for each w implies $A_{i,t}(w)$ is the same across all $t \in \mathbb{T}$; consequently, we can eliminate the t index and denote it simply as

$$\alpha_w = A_i(w) \tag{6.24}$$

The relationship between the random variables $Y_{i,t}$, $X_{i,t}$, A_i, and $\mathcal{E}_{i,t}$ for each $i \in \{1, 3, \ldots, N\}$ is therefore

$$Y_{i,t} = A_i(w) + \beta \cdot X_{i,t}(w) + \mathcal{E}_{i,t}(w) \tag{6.25}$$

Intuitively, these are indexed random variables because we are going to engage a data generating process for each $i \in \{1, \ldots, N\}$ occasions, and for each w we capture from Ω we will take measurements at times $t \in \{1, \ldots, T\}$.

Moreover, if $E(\mathcal{E}_{i,t} | X_{i,t} = x) = 0$ for all x, and $E(A_i | X_i = x) = \alpha$, a constant, for all x such that

$$A_i(w) = \alpha + \Psi_i(w) \tag{6.26}$$

with $E(\Psi_i) = 0$, then we have

$$Y_{i,t} = \alpha + \beta \cdot X_{i,t}(w) + \mathcal{E}_{i,t}{}^*(w) \tag{6.27}$$

in which $\mathcal{E}_{i,t}{}^*(w) = \Psi_i(w) + \mathcal{E}_{i,t}(w)$ and $E(\mathcal{E}_{i,t}{}^* | X_{i,t} = x) = 0$ for all x.

For the convenience of explanation, assume Ψ_i and $\mathcal{E}_{i,t}$ are independent of each other, and assume the conditional variances are the same across values of X. In this case, the variance of $\mathcal{E}_{i,t}{}^*$ is the variance of the sum $\Psi_i + \mathcal{E}_{i,t}$. Because the expected values of Ψ_i and $\mathcal{E}_{i,t}$ are both 0, the variance is

$$\mathrm{Var}(\mathcal{E}_{i,t}^*) = E\left[(\Psi_i(w) + \mathcal{E}_{i,t}(w))^2\right] \tag{6.28}$$

which expanding the square and pushing the expectation through the linear terms yields

$$\mathrm{Var}(\mathcal{E}_{i,t}^*) = E\left[\Psi_i(w)^2\right] + E[2 \cdot \Psi_i(w) \cdot \mathcal{E}_{i,t}(w)] + E\left[\mathcal{E}_{i,t}(w)^2\right] \tag{6.29}$$

Noting that, because Ψ_i and $\mathcal{E}_{i,t}$ are taken to be independent, the expectation of the second right-hand side term of Equation 6.29 is zero, leaving us with the sum of two expectations, each of the square of terms with expectations of zero. The variance on the left-hand side is therefore the sum of two variances:

$$\mathrm{Var}(\mathcal{E}_{i,t}^*) = \mathrm{Var}[\Psi_i(w)] + \mathrm{Var}[\mathcal{E}_{i,t}(w)] \tag{6.30}$$

What is the covariance between $\mathcal{E}_{i,t}^*$ and $\mathcal{E}_{i,t'}^*$, where t and t' are two different arbitrary indices in an index set T? Since both variables are defined for the same observation indexed by i, they will be evaluated at the same element w that is generated by the data generating process. Therefore, the covariance

$$\mathrm{Cov}(\mathcal{E}_{i,t}^*, \mathcal{E}_{i,t'}^*) = E[(\Psi_i(w) + \mathcal{E}_{i,t}(w)) \cdot (\Psi_i(w) + \mathcal{E}_{i,t'}(w))] \tag{6.31}$$

Expanding the square and pushing the expectation through the linear terms of Equation 6.31 yield:

$$\begin{aligned}
\mathrm{Cov}(\mathcal{E}_{i,t}^*, \mathcal{E}_{i,t'}^*) = {} & E[\Psi_i(w) \cdot \Psi_i(w)] + E[\Psi_i(w) \cdot \mathcal{E}_{i,t'}(w)] \ldots \\
& + E[\mathcal{E}_{i,t}(w) \cdot \Psi_i(w)] + E[\mathcal{E}_{i,t}(w) \cdot \mathcal{E}_{i,t'}(w)]
\end{aligned} \tag{6.32}$$

in which the arguments for Ψ and $\mathcal{E}$ are the same w because they are both determined by the same observation i, which will thereby have the same w from the data generating process.

Here is where the notational simplification we achieved by dropping indices is helpful. We immediately see that the first right-hand side term of Equation 6.32 is

$$E[\Psi_i(w) \cdot \Psi_i(w)] = E[\Psi_i(w)^2] = V[\Psi_i] \qquad (6.33)$$

Without this notational reduction, we would have written $E[\Psi_{i,t}(w) \cdot \Psi_{i,t'}(w)]$, and we would have had to remember, notwithstanding the different indices, that the two indicated variables are the same and, therefore, the covariance between them is equal to the variance of one of them.

If $\text{Cov}(\Psi_i, \mathcal{E}_{i,t}) = 0$ and $\text{Cov}(\mathcal{E}_{i,t}, \mathcal{E}_{i,t'}) = 0$ for all $i \in I$ and all $(t, t') \in T \times T$ such that $t \neq t'$, then $\text{Cov}(\mathcal{E}_{i,t}^*, \mathcal{E}_{i,t}^*) = V(\Psi_i)$, that is, the variance of Ψ_i. Otherwise, the covariances among the Ψ_i and $\mathcal{E}_{i,t}$ variables would be included as well.

Next, it is more important to ask what the covariance is between $\mathcal{E}_{i,t}^*$ and $\mathcal{E}_{j,t}^*$ for $(i, j) \in I \times I$ and $i \neq j$. In other words, what is the covariance between observations? In terms of the preceding example, we have

$$\text{Cov}(\mathcal{E}_{i,t}^*, \mathcal{E}_{j,t}^*) = E[(\Psi_i(w) + \mathcal{E}_{i,t}(w)) \cdot (\Psi_j(w') + \mathcal{E}_{j,t}(w'))] \qquad (6.34)$$

Note the use of distinct symbols w and w' in Equation 6.34 to represent the arbitrary elements of Ω associated with the data generating processes indexed by i and j, respectively. The reason for this is that although $\mathcal{E}_{i,t}^*$ and $\mathcal{E}_{j,t}^*$ have the same domain Ω, they represent two different occurrences of the data generating process and thereby the outcome from Ω obtained for observation i is not necessarily the same as the outcome obtained for observation j.

Expanding the square and pushing the expectation operator through the linear terms of Equation 6.34 give

$$\text{Cov}(\mathcal{E}_{i,t}^*, \mathcal{E}_{j,t}^*) = E[(\Psi_i(w) + \mathcal{E}_{i,t}(w)) \cdot (\Psi_j(w') + \mathcal{E}_{i,t}(w'))] \qquad (6.35)$$

which expanding and extending the expectation through the term yields

$$\text{Cov}(\mathcal{E}_{i,t}^*, \mathcal{E}_{j,t}^*) = E[\Psi_i(w) \cdot \Psi_j(w')] + E[\Psi_i(w) \cdot \mathcal{E}_{j,t}(w')] \dots \\ + E[\mathcal{E}_{i,t}(w) \cdot \Psi_j(w')] + E[\mathcal{E}_{i,t}(w) \cdot \mathcal{E}_{j,t}(w')] \qquad (6.36)$$

Consider the first expectation on the right-hand side of Equation 6.36:

$$E[\Psi_i(w) \cdot \Psi_j(w')] = \sum_w \sum_{w'} [\Psi_i(w) \cdot \Psi_j(w')] \cdot P(\{w'\}|\{w\}) \cdot P(\{w\}) \qquad (6.37)$$

However, the data generating process is one such that $P(\{w'\}|\{w\}) = P(\{w'\})$ for all $w \in \Omega$ (i.e., the observations are independent). This yields,

$$E[\Psi_i(w) \cdot \Psi_j(w')] = \sum_w \sum_{w'} [\Psi_i(w) \cdot \Psi_j(w')] \cdot P(\{w'\}) \cdot P(\{w\}) \qquad (6.38)$$

which, grouping the terms for w', is

$$E[\Psi_i(w) \cdot \Psi_j(w')] = \sum_w \Psi_i(w) \underbrace{\left(\sum_{w'} [\Psi_j(w') \cdot P(\{w'\})] \right)}_{A} \cdot P(\{w\}) \quad (6.39)$$

The summation denoted by A in Equation 6.39 is the expectation of Ψ_j, which is 0; consequently, $E[\Psi_i(w) \cdot \Psi_j(w')]$ is 0. Note that we could just as well have considered the expectation of Ψ_i rather than Ψ_j and obtained the same result.

Here it is important to note that the number of summations corresponds to the number of individual arguments, which in this case is two: one for w and one for w'.

The key to this result is that the data generating process was independent regarding the production of observations for i and j. As discussed in Chapter 4, all random variables across independent observations are independent. Consequently, if we were to investigate the remaining expectations of $E[\Psi_i(w) \cdot \mathcal{E}_{j,t}(w')]$, $E[\mathcal{E}_{i,t}(w) \cdot \Psi_j(w')]$, and $E[\mathcal{E}_{i,t}(w) \cdot \mathcal{E}_{j,t}(w')]$, we will obtain the same result for each—each expectation is 0. The full result is then $\text{Cov}(\mathcal{E}_{i,t}^*, \mathcal{E}_{j,t}^*) = 0$ since observations i and j are independent.

Due to the independence of the observations, the preceding example is uninteresting; however, it points to the usefulness of the notational conventions, particularly regarding dropping indices that do not differentiate variables, and regarding the need to use different notations for arbitrary elements of the random variables' domain across variables (e.g., the use of w and w' above). Let us now consider a slightly different data generating process.

Example 6.2

Consider the same structural model used in Example 6.1,

$$Y_{wt} = \alpha_w + \beta \cdot X_{wt}(d) + \varepsilon_{wt} \quad (6.40)$$

but with a different data generating process. This time the data generating process is one in which we randomly select an individual w and then randomly select two times at which observations are made. We measure the response of the individual to each time. The interesting question is how does this same structural model translate to random variables of this data generating process with random assignment of doses?

The probability space representing a sample of size N each having two randomly selected times is $((\Omega \times T)^{N \cdot 2}, \mathcal{A}^{N \cdot 2}, P)$. Each observation, and consequent random variable, of this nested data generating process can be indexed by a single observation number. However, in this case it is helpful to use two indices: one being $i \in \{1, \ldots, N\}$ representing the part of the process that selects an individual, the other being $j \in \{1, 2\}$ representing the two random selections of times at which measurements are made. The component probability space for an observation indexed by ij is then $(\Omega \times T, \mathcal{A}, P_{ij})$. By inspection, it should be easy to see that this indexing scheme uniquely distinguishes observations and corresponding variables, which is the goal of indexing.

Expressing our relationship between the structural variables,

$$Y_{wt} = \alpha_w + \beta \cdot X_{wt}(d) + \varepsilon_{wt} \tag{6.41}$$

in terms of the corresponding random variables for an arbitrary observation yields

$$Y_{ij} = A_{ij}(w) + \beta \cdot X_{ij}(w,t) + \mathcal{E}_{ij}(w,t) \tag{6.42}$$

The random variable Y_{ij} is measured as just that response for individual w at time t for the intervention dose individual w happens to be subject to at time t, which is $Y_{wt}(d)$ of the structural model; a similar connection exists between the random variable $X_{ij}(w, t)$ and its structural counterpart $X_{wt}(d)$. Note that here, unlike the previous example, the indices for the random variables are not separated by a comma. This is because both indices, together, provide the unique designation of the observational process that will select a pair (w, t) from $\Omega \times T$.

The individual and time-specific constants α_w and ε_{wt} in the structural model can vary across individuals and are therefore random variables $A_{ij}(w)$ and $\mathcal{E}_{ij}(w, t)$ when defined in terms of the probability space ($\Omega \times T$, $\mathcal{A}$, P_{ij}). Although the full notation of $A_{ij}(w)$ is $A_{ij}(w, t)$, it does not vary across t and therefore t is dropped as an argument. Similarly, we drop the j index from $A_{ij}(w)$ because the variables are the same across j.

If, as in Equation 6.26, we specify

$$A_i(w) = \alpha + \Psi_i(w) \tag{6.43}$$

with $E(\Psi_i) = 0$, then we have

$$Y_{ij} = \alpha + \beta \cdot X_{ij}(w, d) + \mathcal{E}_{ij}^*(w, t) \tag{6.44}$$

in which $\mathcal{E}_{ij}^*(w, t) = \Psi_i(w) + \mathcal{E}_{ij}(w, t)$.

The interesting question is what are the covariances between observations of this data generating process? I will consider two: $\mathrm{Cov}(Y_{ij}, Y_{ij'})$ and $\mathrm{Cov}(Y_{ij}, Y_{i'j})$. The remaining two combinations can be worked out as easily. For ease of presentation, I will consider the variables A_i, X_{ij}, and $\mathcal{E}_{ij}$ to be mutually independent.

The $\mathrm{Cov}(Y_{ij}, Y_{ij'})$ can be expressed as

$$\mathrm{Cov}(Y_{ij}, Y_{ij'}) = E[(\beta \cdot \dot{X}_{ij}(w, t) + \Psi_i(w) + \mathcal{E}_{ij}(w, t)) \\ \times (\beta \cdot \dot{X}_{ij'}(w, t') \ldots + \Psi_i(w) + \mathcal{E}_{ij'}(w, t'))] \tag{6.45}$$

This is the same as

$$\mathrm{Cov}(Y_{ij}, Y_{ij'}) = \beta^2 \cdot E[\dot{X}_{ij}(w, t) \cdot \dot{X}_{ij'}(w, t')] + E[\Psi_i(w) \cdot \Psi_i(w)] \ldots \\ + E[\mathcal{E}_{ij}(w, t) \cdot \mathcal{E}_{ij'}(w, t')] \tag{6.46}$$

in which $\dot{X}_{ij}$ denotes the appropriate mean-centered variable. Note the careful use of the w, t, and t'. Both Y_{ij} and $Y_{ij'}$ are based on the same i and will therefore have values determined by the same w that is obtained from that part of the data generating process. However, the variables are based on different indices indicating the two different occurrences of randomly selected times and can therefore produce different times t and t'. Therefore, Y_{ij} will not necessarily be the same as $Y_{ij'}$ due to the different times that observations ij and ij' can produce. Consequently, we use t to indicate the arbitrary time for Y_{ij} and t' to indicate the arbitrary time for $Y_{ij'}$. This is a very important notational convention as it will dictate how the probabilities are applied to the random variables.

Each expectation in the preceding equation is as follows. Regarding the first,

$$E[\dot{X}_{ij} \cdot \dot{X}_{ij'}] = \sum_w \sum_t \sum_{t'} \dot{X}_{ij}(w,t) \cdot \dot{X}_{ij'}(w,\ t') \cdot P(\{t'\}|\{w\})$$
$$\cdot P(\{t\}|\{w\}) \cdot P(\{w\}) \tag{6.47}$$

which, grouping the terms under the summation for t', is

$$E[\dot{X}_{ij} \cdot \dot{X}_{ij'}] = \sum_w \sum_t \dot{X}_{ij}(w,t) \cdot \left[\sum_{t'} \dot{X}_{ij'}(w,t') \cdot P(\{t'\}|\{w\}) \right]$$
$$\cdot P(\{t\}|\{w\}) \cdot P(\{w\}) \tag{6.48}$$

Here again it is important to note the number of summations correspond to the number of distinct arguments, which in this case is three: there is no sum over a second w' element because both random variables will take on values associated with the same w that is captured in the data generating process.

Because in equation 6.48 the summation with respect to t' is the expected value of $\dot{X}_{ij}$ conditional on $\{w\}$, and similarly the summation with respect to t is the expected value of $\dot{X}_{ij'}$ conditional on $\{w\}$, however these conditional expectations are the same.

Therefore, equation 4.48 can be expressed in terms of the square of either:

$$E[\dot{X}_{ij} \cdot \dot{X}_{ij'}] = \sum_w E(\dot{X}_{ij}|\{w\})^2 \cdot P(\{w\}) = \sum_w E(\dot{X}_{ij'}|\{w\})^2 \cdot P(\{w\}) \tag{6.49}$$

which is the variance of the conditional expectation plus the expectation of the conditional expectation squared or either random variable $\dot{X}_{ij}$ or $\dot{X}_{ij'}$:

$$E[\dot{X}_{ij} \cdot \dot{X}_{ij'}] = V(E(\dot{X}_{ij}|\{w\})) + E(E(\dot{X}_{ij}|\{w\}))^2$$
$$= V(E(\dot{X}_{ij'}|\{w\})) + E(E(\dot{X}_{ij'}|\{w\}))^2 \tag{6.50}$$

However, since the squared terms in equation 6.50 are equal to 0, the first term on the right-hand side of equation 6.46 is simply the squared parameter β multiplied by the variance of the conditional expectation:

$$\beta^2 \cdot E[\dot{X}_{ij}(w,t) \cdot \dot{X}_{ij'}(w,t')] = \beta^2 \cdot V(E(\dot{X}_{ij}|\{w\})) = \beta^2 \cdot V(E(\dot{X}_{ij}|\{w\})) \tag{6.51}$$

Regarding $E[\Psi_i(w) \cdot \Psi_i(w)]$:

$$E[\Psi_i(w) \cdot \Psi_i(w)] = E\left[\Psi_i(w)^2\right] \tag{6.52}$$

But, given the expected value of $\Psi_i(w)$ is zero, this expectation of its squared value is the variance:

$$E[\Psi_i(w) \cdot \Psi_i(w)] = \text{Var}[\Psi_i(w)] \tag{6.53}$$

and, regarding $E[\mathcal{E}_{ij}(w,t) \cdot \mathcal{E}_{ij'}(w,t')]$:

$$E[\mathcal{E}_{ij}(w,t) \cdot \mathcal{E}_{ij'}(w,t')] = \text{Cov}[\mathcal{E}_{ij}(w,t), \mathcal{E}_{ij'}(w,t')] \tag{6.54}$$

Consequently, regarding random variables across observations that necessarily capture the same person but different times, the covariance is

$$\text{Cov}(Y_{ij}, Y_{ij'}) = \text{Var}[\Psi_i(w)] + \text{Cov}[\mathcal{E}_{ij}(w,t), \mathcal{E}_{ij'}(w,t')] \tag{6.55}$$

However, by the data generating process, $\mathcal{E}_{ij}$ and $\mathcal{E}_{ij'}$ are independent; therefore, $\text{Cov}[\mathcal{E}_{ij}(w,t), \mathcal{E}_{ij'}(w,t')] = 0$ in Equation 6.55 and overall $\text{Cov}(Y_{ij}, Y_{ij'}) = \text{Var}[\Psi_i(w)]$.

Now consider the covariance of random variables across observations that potentially capture different people. The $\text{Cov}(Y_{ij}, Y_{i'j})$ is

$$\text{Cov}(Y_{ij}, Y_{i'j}) = E[(\beta \cdot \dot{X}_{ij}(w,t) + \Psi_i(w) + \mathcal{E}_{ij}(w,t)) \cdot (\beta \cdot \dot{X}_{i'j}(w',t') \ldots$$
$$+ \Psi_{i'}(w') + \mathcal{E}_{i'j}(w',t'))] \tag{6.56}$$

which, after expanding the polynomial and pushing the expectation across the linear terms, yields

$$\text{Cov}(Y_{ij}, Y_{i'j}) = \beta^2 \cdot E[\dot{X}_{ij}(w,t) \cdot \dot{X}_{i'j}(w',t')] + E[\Psi_i(w) \cdot \Psi_{i'}(w')] \ldots$$
$$+ E[\mathcal{E}_{ij}(w,t) \cdot \mathcal{E}_{i'j}(w',t')] \tag{6.57}$$

Each expectation in Equation 6.57 is as follows. Regarding the first expectation,

$$E[\dot{X}_{ij} \cdot \dot{X}_{i'j}] = \sum_w \sum_{w'} \sum_t \sum_{t'} \dot{X}_{ij}(w,t) \cdot \dot{X}_{i'j}(w',t') \cdot P(\{t\}|\{w\})$$
$$\cdot P(\{t'\}|\{w'\}) \cdot P(\{w\}) \cdot P(\{w'\}) \tag{6.58}$$

After grouping the summations for t and t', the expectation is

$$E[\dot{X}_{ij} \cdot \dot{X}_{i'j}] = \sum_w \sum_{w'} \left[\sum_t \dot{X}_{ij}(w,t) \cdot P(\{t\}|\{w\}) \cdot \sum_{t'} \dot{X}_{i'j}(w',t') \cdot P(\{t'\})|\{w'\} \right]$$
$$\cdot P(\{w'\}) \cdot P(\{w\}) \tag{6.59}$$

Since the summations $\sum_t \dot{X}_{ij}(w,t) \cdot P(\{t\}|\{w\})$ and $\sum_{t'} \dot{X}_{i'j}(w',t') \cdot P(\{t'\}|\{w'\})$ are equal to 0, being the expected value of mean-centered variables, the expectation is 0:

$$E[\dot{X}_{ij} \cdot \dot{X}_{i'j}] = 0. \tag{6.60}$$

In this case, there are four summations in Equation 6.59 because these random variables will take on values associated with potentially different elements of Ω, indicated here as w and w', and potentially different elements of T here denoted as t and t'. The selection of individuals and the selection of times are mutually independent; consequently, the probabilities are those associated with independence. The last equality, by which the expectation is equal to 0, is due to the same argument presented above for $\text{Cov}(Y_{ij}, Y_{ij'})$.

This example shows that the number of distinct arguments for random variables do not necessarily follow the number of distinct indices. Consider, for example, the two variables $\dot{X}_{ij}(w, t)$ and $\dot{X}_{i'j}(w', t')$. Both have the same index j, but different associated arguments t and t'. This is because j is indexing the time-selection process that is nested within the person-selection process. If, for example, $j = 1$, then both random variables are associated with the first occurrence of randomly selecting a time, but since these are different random variables due to being associated with different occurrences of randomly selecting individuals (i.e., i and i'), the resultant randomly selected times can be different for each and are consequently denoted differently as t and t'. It would have a different interpretation if the data generating process was not nested, such as randomly selecting individuals and then randomly selecting two times at which all individuals are measured. In that case, we would have $\dot{X}_{ij}(w, t)$ and $\dot{X}_{i'j}(w', t)$ because both observations would be based on the same randomly selected times. Properly interpreting the current notation requires understanding how it relates to the data generating process—notation alone is not sufficient. We could add more nuance to allow the notation to directly indicate nested versus nonnested processes; for example, we could use parentheses to denote nesting, such as $X_{i(j)}$ to denote j nested within i. However, for the purpose of this text, I will forgo greater notational complexity and depend on our understanding of the data generating process that we are modeling.

Returning to our example, regarding the second expectation of Equation 6.57:

$$E[\Psi_i(w) \cdot \Psi_{i'}(w')] = \sum_{w} \sum_{w'} \Psi_i(w) \cdot \Psi_{i'}(w') \cdot P(\{w\}) \cdot P(\{w'\}) \qquad (6.61)$$

which, after rearranging the summations is

$$E[\Psi_i(w) \cdot \Psi_{i'}(w')] = \sum_{w} \Psi_i(w) \cdot P(\{w\}) \cdot \sum_{w'} \Psi_{i'}(w') \cdot P(\{w'\}) \qquad (6.62)$$

and noting that each summation is an expectation equal to 0,

$$E[\Psi_i(w) \cdot \Psi_{i'}(w')] = E[\Psi_i(w)] \cdot E[\Psi_{i'}(w')] \qquad (6.63)$$

the overall expected value is 0, that is, $E[\Psi_i(w) \cdot \Psi_{i'}(w')] = 0$.

Regarding the term $E[\mathcal{E}_{ij}(w, t) \cdot \mathcal{E}_{i'j}(w', t')]$ in Equation 6.57, similar to above, the expectation turns out to be 0 because, again, they are based on independent observations.

Consequently, the covariance between Y_{ij} and $Y_{i'j}$ is 0:

$$\text{Cov}(Y_{ij}, Y_{i'j}) = 0 \qquad (6.64)$$

Examples 6.1 and 6.2 show three important features of the notation used in the remainder of this book: (1) the elimination of those indices and arguments over which a random variable does not vary, (2) distinguishing arguments that correspond to results for different indices (e.g., the use of w and w', and d and d'), and (3) assuring that summation is only taken over those arguments that correspond to the variables under consideration. The indexing notation must be understood based on an understanding of the context of a particular data generating process. Careful attention to these points will allow us to determine the dependence across any pair of random variables.

Random versus Fixed Effects

The terms "random effect," "random coefficient," "random parameter," "hierarchical," and "multilevel" are all commonly (although not exclusively) used to label models of random variables from data generating processes with observations represented by probability spaces such as $(S \times \Omega, \mathcal{A}, P)$ in which elements of Ω are associated by a nonzero P with only one element of S. Assuming our sigma-algebra is the power set, this means that for all elements w of Ω, there is only one element s of S such that $P(\{(s, w)\}) > 0$. In formal logic, assuming the domain of the arguments is understood, this is represented as $\forall w \exists! s (P(\{(s, w)\}) > 0)$, which is to say that for all w's there is exactly one s for which the probability of $\{(s, w)\} \in \mathcal{A}$ is greater than 0. The data generating process is one where an element is sampled from S and then an element is sampled from Ω based on a probability conditional on the result from S. For example, randomly selecting a school from a population of schools (S) and then randomly selecting a student from within the obtained school (the subset of Ω corresponding to student in the selected school), or randomly selecting a physician from a population of physicians (S) and then randomly selecting a patient from the obtained physician's patient panel (the subset of Ω corresponding to the selected physician's patient panel), or randomly selecting an individual (S) and then randomly selecting a time to measure a characteristic (the subset of Ω corresponding to an index set of time). These are examples of nested data generating processes.

It is possible to specify the outcome set with more than two components thereby indicating more elaborate nesting: for example, randomly selecting a hospital from a population of hospitals, and randomly selecting a clinical department within the obtained hospital, then randomly selecting a physician within the selected department, and finally, randomly selecting a patient of the selected physician. Such a process would have an outcome set defined by H × D × P × I in which H denotes the set of hospitals, D denotes the set of clinical departments, P denotes physicians, and I denotes patients. However, for the purposes of this presentation, I will restrict the discussion to the two-level models: more levels are a straightforward extension.

Notwithstanding the above description of a typical random effects model, a more general presentation simply assumes that there is a measurable partition on a product outcome set generated by one of the component sets across which the distribution of random variables under investigation vary. This characteristic alone does not differentiate the partition from other random variables, but as stated above for random and fixed effects models, these partitions typically reflect structure of the data generating process (e.g., clusters in a cluster sampling design). However, as the following sections show, there are other applications as well, each having other reasons to distinguish these partitions from other random variables of interest in our analysis.

Suppose $(S \times \Omega, \mathcal{A}, P)$ is a probability space modeling a data generating process in which S contains K elements. Define $\Pi = \{\pi_1, \pi_2, \ldots, \pi_K\}$ as a measurable partition of $S \times \Omega$ in $\mathcal{A}$ generated by the elements of S. Specifically, an arbitrary element of the partition is defined as $\pi_k = \{(s, w): s = s_k, s \in S, w \in \Omega\}$; consequently, π_k includes all elements that have s equal to s_k. If we have a random variable Y, random vector $\mathbf{X}$ of explanatory variables, and a random vector $\mathbf{Z}$ of indicators of membership in each of the K elements of the partition Π (i.e., for all $(s, w) \in S \times \Omega$, $Z_k((s, w)) = 1((s, w) \in \pi_k)$ for each $k \in \{1, 2, \ldots, K\}$), then the parameter vector $\boldsymbol{\theta} = (\theta_1, \theta_2, \ldots, \theta_K)'$ associated with $\mathbf{Z}$ in the density $f(Y | \mathbf{X}, \mathbf{Z}; \boldsymbol{\theta}, \boldsymbol{\phi})$ comprises what are commonly called *fixed effects*. Note that $\boldsymbol{\phi}$ is a vector of remaining parameters of the distribution.

Rather than specifying a vector of indicators $\mathbf{Z}$, we can just as well directly define $\boldsymbol{\theta}$ as Θ a random variable,

$$\Theta((s, w)) = \begin{cases} \theta_1 & \text{if } (s, w) \in \pi_1 \\ \theta_2 & \text{if } (s, w) \in \pi_2 \\ \vdots & \vdots \\ \theta_K & \text{if } (s, w) \in \pi_K \end{cases} \tag{6.65}$$

and restate the density as $f(Y | \mathbf{X}, \Theta; \boldsymbol{\phi})$. In this case, we consider the random variable Θ as comprising the fixed effects. Note that $\boldsymbol{\phi}$ is not likely to include parameters associated with Θ because Θ likely enters the distribution directly as if it was itself a parameter, which it was in the preceding specification using $\mathbf{Z}$. In this case, if corresponding parameters were included in $\boldsymbol{\phi}$, they would be fixed constants rather than parameters free to be estimated. This second form, that is, including Θ as a random variable, is a common representation for discussing fixed effects and identifying certain types of estimators (e.g., what are commonly called within estimators that ultimately eliminate the fixed effects from the model thereby negating the need to estimate them). The density in terms of $\mathbf{Z}$ is common as a specification for other methods of estimating fixed effects (e.g., methods that include the set of indicators of the partition in the model to be estimated). These two representations are not necessarily identical. Because many elements of the partition may have the same value for Θ, controlling for Θ may be the same as controlling for multiple $\mathbf{Z}$'s.

As a random variable, Θ is called a *random effect* in the joint density $f(Y, \Theta \mid \mathbf{X}; \boldsymbol{\phi})$. In the case where Θ and $\mathbf{X}$ are dependent, then

$$f(Y, \Theta \mid \mathbf{X}; \boldsymbol{\phi}) \, f(Y \mid \mathbf{X}, \Theta; \boldsymbol{\phi}) \cdot f(\Theta \mid \mathbf{X}; \boldsymbol{\varphi}) \tag{6.66}$$

These models are often called *correlated random effects models* because $f(\Theta \mid \mathbf{X}; \boldsymbol{\varphi})$ is taken to be a nontrivial function of $\mathbf{X}$. In the case where Θ and $\mathbf{X}$ are independent, then

$$f(Y, \Theta \mid \mathbf{X}; \boldsymbol{\phi}) = f(Y \mid \mathbf{X}, \Theta; \boldsymbol{\phi}) \cdot f(\Theta \mid \boldsymbol{\varphi}) \tag{6.67}$$

models of this type are typically just called *random effects models*.

Be aware, however, that the labeling of fixed effects and random effects is not consistent across their use; you must be sure that you understand how the terms are specifically being applied when reading any given literature. Fortunately, the underlying characteristics of the relevant distributions are usually the same, notwithstanding the subtle difference in application of terminology.

Perhaps the simplest example of fixed and random effects models is one in which the random variable Y_i is a linear function of random variables Θ_i, $\mathbf{X}_i$, and $\mathcal{E}_i$:

$$Y_i = \Theta_i + \beta \cdot \mathbf{X}_i + \mathcal{E}_i \tag{6.68}$$

Note that here I am using the subscript i as a simple index of the observation and not using a two-index notation that would capture the structure of the data generating process.

If we are interested in the distribution of Y conditional on both random variables X and Θ, then we have a fixed effects model in which the, possibly subset of, elements of Θ that are conditioned on comprise the set of fixed effects. If we are interested in the distribution of Y conditional on X only, then we have a random effects model in which the random variable Θ is a random effect. In both models, the intercept of the linear function is taken to vary across the measurable partition Π and is a random variable—one that we either condition on as a fixed effect or integrate over as a random effect. Note that we could also condition on $\mathbf{Z}$ to obtain a fixed effects model. If the fixed effects associated with each $\mathbf{Z}$ were distinct, then we would have the same model as when conditioning on Θ.

Because Θ is a random variable, conditional distributions are possible and we may express Θ as a function of other variables. For simplicity of exposition, suppose the regression function of Θ is a linear function of measured random variables $\mathbf{W}$. We can model Θ_i as

$$\Theta_i = \varphi \cdot \mathbf{W}_i + \Psi_i \tag{6.69}$$

If we condition on $\mathbf{W}$, the parameters φ are sometimes called fixed effects and Ψ_i is considered a random effect. The corresponding model is a *mixed effect* model. For example, if the partition is indicating hospitals, then the hospital effect Θ_i may be expressed as a function of measured hospital characteristics and a hospital random effect.

Suppose we are interested in two measurable partitions Π_1 and Π_2 defined on an outcome set $S \times Q \times \Omega$ of a probability space $(S \times Q \times \Omega, \mathcal{A}, P)$ for which Π_1 is generated by the elements of S and Π_2 is generated by the elements of Q. If we define corresponding random variables Θ_{i1} and Θ_{i2}, we may condition on each variables (thereby incorporating two fixed effects), or integrate across both (thereby incorporating two random effects), or specify a mixture of the two (thereby obtaining another type of mixed effects model).

Additional constraints apply if one of the partitions is a refinement of the other, for example, Π_2 is a refinement of Π_1 such as indicating physicians (Π_2) within hospital (Π_1), where each physician has privileges at only one hospital. In this case, conditioning on Θ_{i2} fixes the element of Θ_{i1} (e.g., fixing the physician identifies the hospital) and, therefore, we cannot treat Θ_{i2} as a fixed effect while at the same time treating Θ_{i1} as a random effect. However, if $E(\Theta_{i2} \mid \Theta_{i1}) = \overline{\Theta}_{i1}$, a function of S, then we can model Θ_{i2} as

$$\Theta_{i2}(s,q) = \overline{\Theta}_{i1}(s) + (\Theta_{i2}(s,q) - \Theta_{i1}(s,q)) = \overline{\Theta}_{i1}(s) + \Delta_i(s,q) \tag{6.70}$$

in which $\overline{\Theta}_{i1}$ may be treated as a hospital-level fixed effect, and if the deviation Δ is a physician-level random variable, we can treat as a random effect, again obtaining a mixed effect model. In our hospital example, we would specify a hospital-level fixed effect and a physician-level random effect.

It is also common to consider other parameters as varying across the partition: for example, β. In this case, our function may be specified as

$$Y_i = \Theta_i + B_i \cdot X_i + \varepsilon_i \tag{6.71}$$

and B_i is called a *random coefficient* if B_i is not conditioned on. If B_i is conditioned on, then it comprises a type of fixed effects as well. Note that regardless of whether they are treated as fixed or random effects, Θ_i and B_i can be specified as random variables.

Details of the estimation techniques for these models are outside the scope of this chapter; however, the reason I consider these models in this chapter is that they are often misapplied and the interpretations of parameters and estimation of the standard errors are often wrong due to a lack of understanding what the probability space is modeling. Before discussing these fairly common cases that are often mistakenly specified and interpreted, let us consider a typical simple random effects model.

Problem 6.2

You collect an independent random sample of N states from the United States, and then independently randomly sample M residents from within each of those states. Determine the dependence structure of a model of individual characteristic Y as a linear function of characteristic X with an intercept as a state-level random effect.

Solution: This is an example of a classic random effects model with a nested (or clustered) data generating process. The full probability model is $((S \times \Omega)^{N \cdot M}, \mathcal{A}, P)$ in which S is the set of 50 states and Ω is the population of the United States.

We index the data generating process with $\mathbb{I} = \{1, \dots, N\}$ and $\mathbb{K} = \{1, \dots, M\}$ to identify each marginal space $(S \times \Omega, \mathcal{A}, P_{ik})$ and its random variables. The interpretation of $\mathbb{I}$ is the set of observational processes that randomly select states and the interpretation of $\mathbb{K}$ is the set of observational processes that randomly select individuals from within each state. To be more general and allow for different numbers of observations within each state, we can use a set of index sets $\{\mathbb{K}_i : i \in \mathbb{I}\}$; however, for simplicity we will simply refer to $\mathbb{K}$.

The interpretation of $\mathbb{I}$ is not the set of states S, nor is the interpretation of $\mathbb{K}$ the population Ω. The sets S and Ω are component sets of the product $S \times \Omega$, which is the domain for the random variables. Consequently, a random variable may be denoted such as $Y_{ik}(s, w)$ in which i and k distinguish a particular random variable associated with the data generating process's production of a possible outcome (s, w), and the pair (s, w) is an arbitrary element of the random variable's domain $S \times \Omega$ (i.e., the outcome set).

Rather than develop an underlying structural model, for simplicity in this problem I will create a model for the relationship between the random variables directly. To start, for each $(S \times \Omega, \mathcal{A}, P_{ik})$, I define three variables: $Y_{ik}(s, w)$, $X_{ik}(s, w)$, and $\Theta_i(s)$. The first two, Y and X, denote arbitrary characteristics of interest and Θ denotes an indicator of the elements of S (i.e., Θ is a random variable representing a partition in $\mathcal{A}$ generated by the elements of S). Define another random variable that reflects the expected value of Y within each state among those with $X = 0$:

$$A_i(s) = \mathrm{E}(Y_{ik}|\Theta_i(s), X_{ik} = 0) \qquad (6.72)$$

define the regression function for Y as

$$\mathrm{E}(Y_{ik}|\Theta_i, X_{ik}) = A_i(s) + \beta \cdot X_{ik}(s, w) \qquad (6.73)$$

and, define the difference between Y and its expected value as

$$Y_{ik} - \mathrm{E}(Y_{ik}|\Theta_i, X_{ik}) = \mathcal{E}_{ik}(s, w) \qquad (6.74)$$

Overall then, Y can be expressed as

$$Y_{ik} = A_i(s) + \beta \cdot X_{ik}(s, w) + \mathcal{E}_{ik}(s, w) \qquad (6.75)$$

for which $\mathrm{E}(\mathcal{E}_{ik}|\Theta_i, X_{ik}) = 0$.

This is a model in which A_i is a "state-level" random variable reflecting a state's average Y among those with $X = 0$. If we condition on A_i, we have a fixed effects model; if we do not condition on A_i, we have a random effects model. Note that we could replace A_i with a vector of indicators $\mathbf{Z}_{ik}$

comprising dummy variables representing each state and condition on **Z** to obtain another representation of a fixed-effect model. If each state has a distinct value for A_i, these are identical models; if not, these models only slightly vary in that conditioning on A_i may incorporate multiple states.

To evaluate the random effects model in the linear case of Equation 6.75, it is easier to specify $E(A_i) = \alpha$, a constant, and thereby write

$$A_i(s) = \alpha + \Psi_i(s) \tag{6.76}$$

in which $E(\Psi_i) = 0$. We then have

$$Y_{ik} = \alpha + \beta \cdot X_{ik}(s, w) + \Psi_i(s) + \mathcal{E}_{ik}(s, w) \tag{6.77}$$

We can determine the dependence structure of random variables from this specification. For ease of presentation, we will consider the relationships conditional on X assuming Ψ_i and $\mathcal{E}_{ik}$ are independent of X.

First, let us consider the conditional variance of Y_{ik}:

$$V(Y_{ik} \mid X_{ik}) = V(\Psi_i + \mathcal{E}_{ik}) = E\big((\Psi_i(s) + \mathcal{E}_{ik}(s, w))^2\big) \tag{6.78}$$

Expanding the polynomial and pushing the expectation operator through the sum yields

$$E\big((\Psi_i + \mathcal{E}_{ik})^2\big) = E\big(\Psi_i^2\big) + 2 \cdot E(\Psi_i \cdot \mathcal{E}_{ik}) + E\big(\mathcal{E}_{ik}^2\big) \tag{6.79}$$

which is

$$E\big((\Psi_i + \mathcal{E}_{ik})^2\big) = V(\Psi_i) + 2 \cdot \text{Cov}(\Psi_i, \mathcal{E}_{ik}) + V(\mathcal{E}_{ik}) \tag{6.80}$$

Equation 6.80 is the variance of the random effect plus the variance of the individual error term plus twice the covariance between them. If the covariance is 0, which is to say the $E(\mathcal{E}_{ik} \mid A_i, X_{ik}) = 0$ for all values of A_i and X_{ik}, as we have specified, then we have

$$V(Y_{ik} \mid X_{ik}) = V(\Psi_i) + V(\mathcal{E}_{ik}) \tag{6.81}$$

The two variances that compose $V(Y_{ik} \mid X_{ik})$ reflect variation due to sampling states $V(\Psi_i)$ and variation due to sampling individuals within state $V(\mathcal{E}_{ik})$.

Next let us consider the conditional covariance of Y_{ik} and $Y_{ik'}$, which represent random variables based on the same selection of state (thereby both indexed by the same element of $\mathbb{I}$, specifically i) but different selection of individual within state (thereby each being indexed by different elements of $\mathbb{K}$, specifically k and k'). To provide greater concision in the presentation, I will denote $\text{Cov}(Y_{ik}, Y_{ik'} \mid X_{ik}, X_{ik'})$ as $\text{Cov}(Y_{ik}, Y_{ik'} \mid X)$. The covariance is

$$\text{Cov}(Y_{ik}, Y_{ik'} \mid X) = E\big((\Psi_i(s) + \mathcal{E}_{ik}(s, w)) \cdot (\Psi_i(s) + \mathcal{E}_{ik'}(s, w'))\big) \tag{6.82}$$

Here, it is important to note that random variables with indices that differ will also have different corresponding arguments. Consequently, because $\mathcal{E}_{ik}$

and $\mathcal{E}_{ik'}$ differ in their $\mathbb{K}$ index, they also differ in their corresponding argument associated with Ω: $\mathcal{E}_{ik}(s, w)$ and $\mathcal{E}_{ik'}(s, w')$. This represents the fact that values of both variables will be determined by the same randomly selected state s but potentially different randomly selected individuals w and w'. It is this nuance in notation that will allow us to easily determine dependence.

Expanding the product on the right-hand side of the preceding equation and pushing the expectation through the consequent sum yields:

$$\mathrm{E}((\Psi_i + \mathcal{E}_{ik}) \cdot (\Psi_i + \mathcal{E}_{ik'})) = \mathrm{E}(\Psi_i^2(s)) + \mathrm{E}(\Psi_i(s) \cdot \mathcal{E}_{ik'}(s, w')) + \ldots + \mathrm{E}(\mathcal{E}_{ik}(s, w)$$
$$\times \Psi_i(s)) + \mathrm{E}(\mathcal{E}_{ik}(s, w) \cdot \mathcal{E}_{ik'}(s, w')) \tag{6.83}$$

The first expectation on the right-hand side is the variance of the random effect: $\mathrm{V}(\Psi_i)$. The next two expectations are the covariance terms between the random effect and the individual errors. The last term is the variance of the errors. Considering the first covariance term in Equation 6.83,

$$\mathrm{E}(\Psi_i(s) \cdot \mathcal{E}_{ik'}(s, w')) = \sum_s \sum_{w'} (\Psi_i(s) \cdot \mathcal{E}_{ik'}(s, w')) \cdot P(\{w'\} | \{s\}) \cdot P(\{s\}) \tag{6.84}$$

which, after rearranging the summations, is

$$\mathrm{E}(\Psi_i(s) \cdot \mathcal{E}_{ik'}(s, w')) = \sum_s \Psi_i(s) \left[\sum_{w'} \mathcal{E}_{ik'}(s, w') \cdot P(\{w'\} | \{s\}) \right] \cdot P(\{s\}) \tag{6.85}$$

Noting that the summation on w' is itself an expectation yields

$$\mathrm{E}(\Psi_i(s) \cdot \mathcal{E}_{ik'}(s, w')) = \sum_G \Psi_i(s) \cdot \mathrm{E}(\mathcal{E}_{ik'} | \Theta_i(s), X_{ik'}) \cdot P(\{s\}) \tag{6.86}$$

Therefore, the left-hand side of Equation 6.86 is the covariance

$$\mathrm{E}(\Psi_i(s) \cdot \mathcal{E}_{ik'}(s, w')) = \mathrm{Cov}(\Psi_i, \mathcal{E}_{ik'}) \tag{6.87}$$

By a similar argument,

$$\mathrm{E}(\mathcal{E}_{ik}(s, w) \cdot \Psi_i(s)) = \mathrm{Cov}(\mathcal{E}_{ik}, \Psi_i) \tag{6.88}$$

Finally, regarding the last term in Equation 6.83, $\mathrm{E}(\mathcal{E}_{ik}(s, w) \cdot \mathcal{E}_{ik'}(s, w'))$:

$$\mathrm{E}[\mathcal{E}_{ik} \cdot \mathcal{E}_{ik'}] = \sum_s \sum_w \sum_{w'} \mathcal{E}_{ik}(s, w) \cdot \mathcal{E}_{ik'}(s, w') \cdot P(\{w\} | \{s\}, \{w'\})$$
$$\times P(\{w'\} | \{s\}) \cdot P(\{s\}) \tag{6.89}$$

Noting that the probability of events $\{w\}$ conditional on events $\{s\}$ is independent of $\{w'\}$, the expectation is

$$\mathrm{E}[\mathcal{E}_{ik} \cdot \mathcal{E}_{ik'}] = \sum_s \sum_w \sum_{w'} \mathcal{E}_{ik}(s, w) \cdot \mathcal{E}_{ik'}(s, w') \cdot P(\{w\} | \{s\})$$
$$\times P(\{w'\} | \{s\}) \cdot P(\{s\}) \tag{6.90}$$

which is

$$E[\mathcal{E}_{ik} \cdot \mathcal{E}_{ik'}] = \sum_s E(\mathcal{E}_{ik}|\Theta_i(s), X_{ik}) \cdot E(\mathcal{E}_{ik'}|\Theta_i(s), X_{ik'}) \cdot P(\{s\}) \qquad (6.91)$$

the covariance

$$E[\mathcal{E}_{ik} \cdot \mathcal{E}_{ik'}] = \text{Cov}(\mathcal{E}_{ik}, \mathcal{E}_{ik'}) \qquad (6.92)$$

However, the data generating process and model specification is such that $E(\mathcal{E}_{ik}|\Theta_i, X_{ik}) = 0$ and $E(\mathcal{E}_{ik'}|\Theta_i, X_{ik'}) = 0$; consequently, the covariances $\text{Cov}(\Psi_i, \mathcal{E}_{ik'})$, $\text{Cov}(\mathcal{E}_{ik}, \Psi_i)$, and $\text{Cov}(\mathcal{E}_{ik}, \mathcal{E}_{ik'})$ are each equal to 0. Overall, then

$$\text{Cov}(Y_{ik}, Y_{ik'}|X) = V(\Psi_i) \qquad (6.93)$$

This result implies that the covariance between random variables representing the independent selection of individuals within a randomly selected state is the variance of the state-level random effect.

Next, let us consider the conditional covariance of Y_{ik} and $Y_{jk'}$, which represent random variables based on different selections of states (thereby each being indexed by different elements of $\mathbb{I}$, specifically i and j) and different selection of individual within state (thereby each being indexed by different elements of $\mathbb{K}$, specifically k and k'). We have

$$\text{Cov}(Y_{ik}, Y_{jk'}|X) = E[(\Psi_i(s) + \mathcal{E}_{ik}(s,w)) \cdot (\Psi_j(s') + \mathcal{E}_{jk'}(s',w'))] \qquad (6.94)$$

which, upon expanding, is

$$\text{Cov}(Y_{ik}, Y_{jk'}|X) = E(\Psi_i(s) \cdot \Psi_j(s')) + E(\Psi_i(s) \cdot \mathcal{E}_{jk'}(s',w')) \cdots \\ + E(\mathcal{E}_{ik}(s,w) \cdot \Psi_j(s')) + E(\mathcal{E}_{ik}(s,w) \cdot \mathcal{E}_{jk'}(s',w')) \qquad (6.95)$$

Although, as above, we can evaluate each of the expectations on the right-hand side, since ik and jk' index completely independent observations, we know from Chapter 4 that all random variables defined on the marginal space $(S \times \Omega, \mathcal{A}, P_{ik})$ are independent of all random variables defined on $(S \times \Omega, \mathcal{A}, P_{jk'})$ for all $i \neq j$. Consequently, each of the expectations on the right-hand side of Equation 6.95 are 0 and, therefore, the covariance $\text{Cov}(Y_{ik}, Y_{jk'}|X)$ is also 0.

In summary, for the classic nested data generating process described in the problem statement, we can model the intercept as a state-level random effect, which will lead to a conditional variance of the dependent variable equal to the variance of the state-level random effect (Ψ_i) plus the variance of the resident-level error term ($\mathcal{E}_{ik}$). Random variables will be dependent for observations indexed by the same state-level random selection process, having covariance equal to $V(\Psi_i)$, which could be 0, and independent otherwise.

The preceding problem was perhaps the most straightforward implementation of a random effects model to a classic situation. What happens, however, if we consider a different, but also common, situation; one that reflects a slight change in the data generating process. Instead of randomly sampling the state, suppose we use a designated set of states, what then?

Problem 6.3

You collect an independent random sample of N residents from each state in the United States. What is the dependence structure of a model of individual characteristic y as a linear function of characteristic x with an intercept as a state-level random effect modeled as a sampling distribution?

Solution: Suppose we are interested in the relationship between characteristics Y and X on individuals w in the population of people Ω across the 50 states; moreover, we consider for an arbitrary individual w from Ω that the relationship in the following Equation 6.96 holds:

$$Y_w = \alpha_w + \beta \cdot X_w \tag{6.96}$$

By this, we mean that each individual w's level of Y is dictated by the level of some variable X. For notational simplicity in this model, I use Ω as an index set for itself, that is, $\mathcal{J}^{-1}: \Omega \to \Omega$. The model implies the slope is the same for everyone, but there is an individual intercept. This constitutes our structural model of individual responses. Note again that the w is indexing an arbitrary element of the population Ω and not an observation of a data generating process.

For comparison with the preceding example, we specify our probability model as follows: let S denote the set of 50 states and Ω the population of the United States; our outcome set for each observation is then $S \times \Omega$, and our set of events $\mathcal{A}$ is the power set of the outcome set. However, and importantly, the data generating process strongly constricts the probabilities on this measurable space. Specifically, each observation is associated with an explicit state; in other words, because we set out to get a sample from each state, the probability of an observation coming from a given state is 1 for that state and 0 for the others. This will have serious consequences in our analysis of dependence for the related random variables and the reason the following results differ from the preceding problem. Our overall probability model is then $((S \times \Omega)^{50 \cdot N}, \mathcal{A}^{50 \cdot N}, P)$, in which N is denoting the sample size within each of the 50 states.

We index the data generating process with $\mathbb{I} = \{1, \dots, 50\}$ and $\mathbb{K} = \{1, \dots, N\}$ to identify each marginal space $(S \times \Omega, \mathcal{A}, P_{ik})$ and its random variables. The interpretation of $\mathbb{I}$ is the set of states in the United States and the interpretation of $\mathbb{K}$ is the set of observational processes that randomly select individuals from the indexed state. As before, we could generalize to a different index set for the observational processes within each state such as $\{\mathbb{K}_i : i \in \mathbb{I}\}$ to be more specific and allow different numbers of observations within each state; however, to minimize notation, we will indicate only one $\mathbb{K}$.

As in the preceding problem, for each $(S \times \Omega, \mathcal{A}, P_{ik})$, I define three variables: $Y_{ik}(s, w)$, $X_{ik}(s, w)$, and $\Theta_i(s)$. The first two, Y and X, denote arbitrary characteristics of interest and Θ denotes an indicator of the elements of S (i.e., Θ is a random variable representing a partition in $\mathcal{A}$ generated by the elements of S, the states). The model specification associated with this data generating process is identical to that of Problem 6.2.

As above, we define another random variable that reflects the expected value of Y within each state among those with $X = 0$ as

$$A_i(\mathbf{s}) = \mathrm{E}(Y_{ik}|\Theta_i(s), X_{ik} = 0) \tag{6.97}$$

we define the regression function for Y as

$$\mathrm{E}(Y_{ik}|\Theta_i, X_{ik}) = A_i(s) + \beta \cdot X_{ik}(s, w) \tag{6.98}$$

and we the difference between Y and its expected value as

$$Y_{ik} - \mathrm{E}(Y_{ik}|\Theta_i, X_{ik}) = \mathcal{E}_{ik}(s, w) \tag{6.99}$$

Overall then, Y can be expressed as

$$Y_{ik} = A_i(s) + \beta \cdot X_{ik}(s, w) + \mathcal{E}_{ik}(s, w) \tag{6.100}$$

which by definition has $\mathrm{E}(\mathcal{E}_{ik}|\Theta_i, X_{ik}) = 0$.

Here is where we depart from the development in Problem 6.2. The expected value of A_i can be different across indices: in this case, the expectation

$$\mathrm{E}(A_i) = \sum_s A_i(s) \cdot P_{ik}(\{s\}) \tag{6.101}$$

has $P_{ik}(\{s\}) = 1$ if $s = \mathcal{J}(i)$ and 0 if not. This is because, for this data generating process in which we set out to obtain a sample from within each state (i.e., the states are not sampled), each observation has a determined state associated with it. Consequently, each observation is for a specific state, with probability 1. Therefore, all elements of the preceding summation are 0, except for the case in which $s = \mathcal{J}(i)$. Given s is a value of the function $\mathcal{J}(i)$, we can denote the specific s associated with a particular i as s_i:

$$\mathrm{E}(A_i) = A_i(s_i) \tag{6.102}$$

For ease of presentation, we will denote each of these expectations simply as α_i. Unlike above, the expected value of A_i is not the same across each observation; consequently, A_i cannot nontrivially be expressed as $A_i(s) = \alpha + \Psi_i(s)$, but must remain as $A_i(s_i) = \alpha_i$. By substitution, we then have

$$Y_{ik} = \alpha_i + \beta \cdot X_{ik}(s, w) + \mathcal{E}_{ik}(s, w) \tag{6.103}$$

In determining the dependence structure of random variables from this specification, as in the preceding example, we will consider the relationships conditional on X. First, let us consider the conditional variance of Y_{ik}:

$$V(Y_{ik}|X_{ik}) = E\left[((\alpha_i + \mathcal{E}_{ik}) - (E(\alpha_i) + E(\mathcal{E}_{ik})))^2\right] \tag{6.104}$$

However, since the expected value of α_i is just α_i, and the expected value of $\mathcal{E}_{ik}$ is 0, the conditional variance is

$$V(Y_{ik}|X_{ik}) = E\left[(\alpha_i - \alpha_i) + \mathcal{E}_{ik})^2\right] \tag{6.105}$$

which is simply the variance of $\mathcal{E}_{ik}$:

$$V(Y_{ik}|X_{ik}) = E\left[\mathcal{E}_{ik}^2\right] = V(\mathcal{E}_{ik}) \tag{6.106}$$

Unlike the preceding nested data generating process, which had a variance of $V(Y_{ik}|X_{ik}) = V(\Psi_i) + V(\mathcal{E}_{ik})$, this data generating process, which selects a sample from a fixed set of states, has a variance of $V(Y_{ik}|X_{ik}) = V(\mathcal{E}_{ik})$. The variance is not the same for this data generating process, in which we did not sample states, than for the preceding nested data generating process, in which we did sample states.

Next, let us consider the conditional covariance of Y_{ik} and $Y_{ik'}$, which represent random variables based on the same state (thereby both indexed by the same element i of $\mathbb{I}$) but different selection of individual within state (thereby each being indexed by different elements of $\mathbb{K}$, specifically k and k'):

$$Cov(Y_{ik}, Y_{ik'}|X) = Cov((\alpha_i(s) + \mathcal{E}_{ik}(s, w)), (\alpha_i(s) + \mathcal{E}_{ik'}(s, w'))) \tag{6.107}$$

which is simply

$$Cov(Y_{ik}, Y_{ik'}|X) = E(\mathcal{E}_{ik}(s, w) \cdot \mathcal{E}_{ik'}(s, w')) \tag{6.108}$$

a covariance term equal to 0:

$$Cov(Y_{ik}, Y_{ik'}|X) = Cov(\mathcal{E}_{ik}(s, w), \mathcal{E}_{ik'}(s, w')) = 0 \tag{6.109}$$

As with the previous analysis, the $\alpha_i(s)$ term drops out of the covariance because $E(\alpha_i(s)) = \alpha_i(s)$ and, therefore, $\alpha_i(s) - E(\alpha_i(s)) = 0$. Moreover, the covariance $Cov(\mathcal{E}_{ik}(s, w), \mathcal{E}_{ik'}(s, w')) = 0$ for this data generating process, in which independent samples are taken within each state. Consequently, $Cov(Y_{ik}, Y_{ik'}|X) = 0$. Again note that this result is different from that of the preceding nested data generating process, which had a covariance of $Cov(Y_{ik}, Y_{ik'}|X) = V(\Psi_i)$. Within-state random variables are not correlated in the current case, but are correlated in the case of a nested data generating process (if A_i varies across states, otherwise $V(\Psi_i)$ would be 0 in the nested case as well).

Next, let us consider the conditional covariance of Y_{ik} and $Y_{jk'}$, which represent random variables based on different states (thereby each being indexed by different element of $\mathbb{I}$, specifically i and j) and different selection of individual within state (thereby each being indexed by different elements of $\mathbb{K}$, specifically k and k'). Rather than work out the details, I will appeal to our understanding from Chapter 4 that random variables of independent observations are independent; consequently, since the sampling of individuals from within different states are independent, the covariance between variables is $\text{Cov}(Y_{ik}, Y_{jk'}|X) = 0$. This is the same as in the case with a nested data generating process.

In summary, for independent samples of people taken from each of a fixed set of states, the random effect variables are a constant and thereby have a variance of 0. This leads to a variance of $V(Y_{ik}|X_{ik}) = V(\mathcal{E}_{ik})$, a covariance of $\text{Cov}(Y_{ik}, Y_{ik'}|X) = 0$, and $\text{Cov}(Y_{ik}, Y_{jk'}|X) = 0$. These results are different from those of the classic random effects model for nested data sets in Problem 6.2.

What happens if we treat the intercept in Problem 6.3 as a random effect reflecting sampling variation across states as if the data generating process were nested (i.e., as if the states were randomly selected, which is what is computationally assumed for a typical random effect model in many statistical analysis software)? Well, we now incorporate variation across the means of the states into our standard errors as if we were randomly sampling states, incorrectly using an estimate of $V(Y_{ik}|X_{ik}) = V(\Psi_i) + V(\mathcal{E}_{ik})$ rather than the correct $V(Y_{ik}|X_{ik}) = V(\mathcal{E}_{ik})$. And, we would mistakenly treat within-state observations as if they were potentially dependent, incorrectly using $\text{Cov}(Y_{ik}, Y_{ik'}|X) = V(\Psi_i)$ rather than the correct $\text{Cov}(Y_{ik}, Y_{ik'}|X) = 0$. This would be a mistake as we are not sampling states. Why would we care to pretend our estimates could vary according to a process that we did not engage?

Problem 6.4

You collect a random sample of States, and then sample residents from each of those states. Can you legitimately specify a model of individual characteristic y as a linear function of characteristic x with an intercept as a State-specific *fixed effect*?

Solution: Yes. You can always focus on conditional distributions; therefore, you can use a model that conditions on the given states or set of fixed effects.

The preceding problems involve data generating processes that either sample states or define sampling within states. In both cases, we let the data generating process be our guide. If we forget this, however, and let the data be our guide, we can be led into a mistaken approach. This can occur if we mistake data for random variables, and we therefore base our probability model on the structure of data rather than the structure of the data generating process. Consider the following problem.

Problem 6.5

You are given a data set of N records generated by an independent random sample of people who live in the United States. The data include an indicator of the state in which each person resides. Based on this data structure, can you legitimately analyze the data as if the data generating process were clustered by State?

Solution: It is common for people to face such a data set and think of people being nested (or clustered) within state and to use an analytical strategy that accounts for such "nesting." Results will make the assumptions about the variances and dependence among the random variables as addressed in Problem 6.2. This is, however, mistaken.

To capture the notion of clustering by state, suppose we use the measureable space that we used in Problem 6.2: $(S \times \Omega, \mathcal{A})$. The outcome set comprises the pairs (s, w), representing a state and person combination. If we modeled the probabilities according to a nested design, we would have the same probability space as in Problem 6.2 and the corresponding results. However, if we instead model the actual data generating process, we would have $(\Omega, \mathcal{A}, P_i)$ as the probability space for each observation i in the index set I having an interpretation of the set of observations to be obtained $\{1, \ldots, N\}$. The actual outcome set for this data generating process that samples people from the United States does not include sampling states. The data generating process is not nested.

Each person in Ω is a resident of some state, and therefore we can define an indicator of each possible state of residence as a random variable on the probability space $(\Omega, \mathcal{A}, P_i)$; however, random variables do not imply the structure of the data generating process. State of residence is a random variable just as are age, income, political party, and so on. We should no more perform an analysis clustered by state than we would cluster by age, income, or other random variable.

State of residence is irrelevant to defining the probabilities of this data generating process. The corresponding distribution of random variables is determined by the independent random sampling of people having sampling probabilities defined by nationwide independent sampling. This example highlights the need to carefully consider the proper definition of the probability space underlying the random variables that generate data.

This mistake can easily arise when researchers are analyzing data that they did not collect, particularly observational data from natural data generating processes. For example, when a researcher is provided a data set for analysis; observes many patient records with the same physician or many student records with the same school, or any other similar "nesting" structure; and considers that physicians would drive some patient responses or schools would drive student responses, he mistakenly engages a strategy of analysis for a nested data generating process. Here the researcher is mistaking data for random variables and/or random variables for the data generating process that define their underlying probabilities: as has been stated throughout the book, they are not the same. When taking a Frequentist approach to modeling random variables, it is the data generating process that makes the variables random and not the data. The phrase "nested data" is not statistically meaningful in this context.

The preceding problems pose questions about how to properly account for the data generating process when using probability to model this source of uncertainty. The next problem poses a question regarding the scientific merit of always modeling the full data generating process as opposed to ignoring at least part of it.

Problem 6.6

You run a large multisite clinical trial of patients, in which you are able to recruit 100 clinics to participate. Can you treat clinic as a random effect? *Should* you treat clinic as a random effect?

Solution: The answer to the first question is, arguably, yes. You can treat the fact that your recruitment strategy had a less than perfect chance of convincing any given clinic to participate so that if you had repeated the process you might have obtained a different set of willing clinics. By this argument, you might model the uncertainty in the clinic participation generating process with a probability measure and proceed with a random effects analysis as in Problem 6.2. This is a common approach for multisite trials with a large number of sites.

However, *should* you use such a random effects analysis? The answer is, arguably, no. This is a question of scientific utility and not one of mathematical probability. First, suppose that you were to use a clinic-specific fixed effects model. In this case, you can obtain legitimate estimates conditional on the given set of clinics. Generalizability to the overall population of clinics cannot be addressed by statistical arguments and must be based on arguments about clinic homogeneity. The standard errors are interpreted in accordance with the sampling variation due to patients that show up at the clinics. This allows us to understand how the estimates could vary due to the people we happen to have obtained in each clinic. Next, suppose you use a random effects model to account for the sampling of clinics. Unfortunately, due to the idiosyncratic selection process, the parameters of the probability measure modeling the researchers' ability to recruit clinics are not likely identified with the parameters of an interpretable population model of the clinics. Consequently, in the absence of knowing (or having good estimates of) the recruitment probabilities of clinics, generalizability to the overall population of clinics still cannot be addressed by statistical arguments and again must be based on appeals to clinic homogeneity. However, now the standard errors incorporate variation due to the idiosyncratic clinic recruitment strategy, which makes them much less clear in their interpretation. In this case, the random effects model has no scientific advantage over the fixed effects model (i.e., it has no greater statistical generalizability), yet muddles the interpretation of the standard errors.

Inherent Fixed Units, Fixed Effects, and Standard Errors

In this section, we consider the combination of the preceding two sections. We look at how understanding the underlying probability space can help us understand what standard errors we wish to calculate.

Problem 6.7

You have a sample of residents from each of the 62 counties in New York State of the United States. You are interested in the linear relationship across counties of the counties' proportions of hospitalizations and the counties' proportions of poor. You use a least-squared errors algorithm to estimate the quantity of interest. What is the standard error?

Solution: Here the counties are inherently fixed: the counties are what they are, and your data generating process was to get a sample of records from each county—there is no sampling of counties. Using the data from each county, you calculate the proportion of hospitalizations (y) and the proportion of poor (x) for each county. You are interested in $b = (X'X)^{-1}(X'y)$, the OLS estimator (X has a row for each of the 62 counties, and columns for x and a constant). You calculate the linear regression parameter b of y on X by ordinary least squares in your favorite statistical package. How would you interpret the reported standard error of b?

The usual reported standard errors are meaningless! The statistical program assumes that your counties are from a probability sample and report standard errors that use the variation in estimates across counties to calculate standard errors for b associated with a process of sampling counties: a process that did not generate the data and therefore did not produce sampling variation in the estimator.

If you had the true population values for y and x for each county, then b would be *the* linear relationship across counties without variation. So, in the case using samples of individuals within each county, is there a standard error associated with b? Yes. Both y and x for each county are estimated by the samples taken from within each county; different samples within county would produce different y and x values.

We can see this by tracking the probabilities that underlie the distribution of the estimator b. These are a set of probabilities as shown in Figure 6.1. Line 1 of Figure 6.1 presents the individual data generating processes for each of the 62 counties. The outcome set for each county comprises the product of the county population (Ω_s for arbitrary county s) N_s times: for example, $\Omega_s^{N_s} = \Omega_s \times \ldots \times \Omega_s$ in which there are N_s replicates of Ω_s, one for each observation in the sample.

Line 2 of Figure 6.1 shows the two random variables, proportion of hospitalizations and proportion of poor, defined on each of these county-specific probability spaces. Each of these random variables presents a

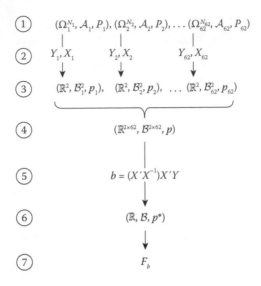

FIGURE 6.1
Tracking probabilities from the initial data generating process to the distribution of the least squared errors estimator.

proportion across the set of N_s individuals that can be captured in the data generating processes. Hence, Y_1 is the proportion of the N_1 individuals from county indexed as 1 who are hospitalized, and X_1 is the corresponding proportion of individuals who are poor. Each pair of random variables map their domain to a subset of the real plane ($\mathbb{R}^2$), specifically to the unit square—see line 3 of Figure 6.1. The probability measure, p_s, that defines the joint distribution of each pair is derived from the probability representing the data generating process of each county (i.e., the P_s in line 1); therefore, the variation in these variables derives from the process of sampling individuals from within each county.

All probability spaces representing the pairs of county-specific random variables shown in line 3 can be combined into the single probability space shown in line 4. This is a product space reflecting the $62 \times 2 = 124$ random variables. Assuming independence of variables across counties, the probability p is the simple product of the p_s probabilities. The distributions of the random variables, such as b in line 5, defined on this space are defined in terms of p, which is derived from the p_s measures, which are in turn derived from the P_s probabilities reflecting sampling. Therefore, as before, the variation in such random variables derives from the original data generating process. The estimator b is a random variable defined on the probability space of line 4. It has a probability measure p^* in line 6 that underlies its sampling distribution F_b in line 7. The variation in the distribution F_b derives from p^*, which is derived from p, which ultimately is derived from the original 62 probabilities P_s.

Consequently, the standard error of b is a function of the within-county data generating processes. Note that the number of counties is irrelevant to the standard error—the logic represented in Figure 6.1 is not changed by considering an arbitrary number of counties (so long as b as defined can be calculated).

The proper standard error for b is the variance associated with $(X'X)^{-1}(X'y)$ in which each measurement of y and x is taken to vary by the data generating process of observed individuals within each county. Because it makes little sense to condition on the observed x values, which are themselves estimates, one would probably calculate the full variance of b and not the variance of b conditional on X.

Note that this has serious ramifications regarding the sample size and power: the number of counties is irrelevant, except to identify the functional relationship (e.g., we would need at least 2 counties for a linear relationship). The number of counties does not impact our sample size considerations. It is the number of individuals sampled within each county that dictates the precision of the county's estimate and thereby impacts the standard error of b. As the sample sizes of individuals within counties go to infinity for all counties, the standard error of b goes to zero, whether you have 2 counties or 62 counties.

The preceding problem addresses using aggregated data within fixed units to study those fixed units. A more common situation is when a smaller unit of analysis is the target of study and samples are obtained from a set of fixed larger units. An example is expressed in the following problem.

Problem 6.8

You have an equal probability sample of residents (with replacement) from each of the 62 counties in New York. You are interested in the relationship between individual-level variables (i.e., measurements on the people of each county). Can you combine the data? What are the standard errors?

Solution: This problem is essentially same as Problem 6.3; however, in that problem I skipped over an important issue, which will be addressed here. Specifically, you have 62 separate data generating processes—one for each county. You can consider a measurable space associated with each county c from the set C of 62 counties as $(\Omega_c, \mathcal{A}_c)$, that is, you would define the 62 measurable spaces $(\Omega_1, \mathcal{A}_1)$, $(\Omega_2, \mathcal{A}_2)$, ..., $(\Omega_{62}, \mathcal{A}_{62})$ that each one represents the residents of a particular county. However, since probability measures can assign nonempty sets the value of zero, you can more simply use a common measurable space $(\Omega, \mathcal{A})$ in which Ω denotes the set of New York State residents and $\mathcal{A}$ is an appropriately rich sigma-algebra (let's just say the power set for convenience). The different data generating processes of the different counties are indicated by their probability measures

P_c which assigns nonzero sampling probabilities to residents of county c and assigns zero to residents of other counties. Your probability spaces are then $(\Omega, \mathcal{A}, P_c)$ for each county $c \in C$.

Because each county has a different probability space, without further assumptions, there is no *a priori* reason to assume that the random variables with corresponding probability spaces $(\mathbb{R}, \mathcal{B}, P_c)$ have the same distributions F_c or shared characteristics (e.g., a regression function). Consequently, without further assumptions, there is no *a priori* reason to believe that $F_c = F_d$ (or any shared distribution characteristics) for any arbitrary pairs of counties c and d; therefore, there is no reason to believe that data from one county provide information about the distribution of random variables for any other county, and the data from the different counties cannot be combined in analysis!

Suppose, however, that it is reasonable to assume the distribution of random variables for each county is of the same parametric family and differs across counties solely by their parameterization. Then you can write each distribution F_c as F_{θ_c}, a distribution indexed by a county-specific parameter θ_c rather than merely by the county, thereby indicating that the distribution varies only in its parameterization and not in its family. If all parameters are distinct across counties, which some would call a fully interacted model, you have gained very little by this reduction because your analysis would be the same as assessing each county's data separately. On the contrary, if some of the parameters in θ are considered to be the same across counties, you are in a position to use the combined data to inform your estimates.

To make this point clear, suppose you are interested in the conditional distribution of random variable Y given values of random variable X, and further suppose it is reasonable to model this distribution as being a member of the normal family of distributions in which the mean is a linear function of X. Then for each county, $Y_c \sim N(\alpha_c + \beta_c \cdot X, \sigma_c^2)$. As stated above, without further assumptions we have gained little in terms of combining the data, but if it is plausible to assume the conditional variance is the same across counties, then $\sigma_c^2 = \sigma^2$ for all counties c, and $Y_c \sim N(\alpha_c + \beta_c \cdot X, \sigma^2)$. In this case, you can combine the data and estimate the county-specific parameters α_c and β_c (by the judicious use of dummy variables as county indicators—assuming you have a large enough sample size) and the common parameter σ^2. This will provide a more precise estimate of σ^2 (since one can use all data to inform this parameter), which will improve your standard error estimators for the remaining parameters.

Suppose it is also reasonable to assume $\beta_c = \beta$ for all counties c. Then our distribution becomes $Y_c \sim N(\alpha_c + \beta \cdot X, \sigma^2)$ and you have the classic fixed effects model in which the intercepts α_c are allowed to vary across counties and can be estimated by combining the data and including county dummy variables in the model, and again greater efficiency is achieved by combining the data.

Other restrictions are available as well: for example, it might be reasonable to also assume $\alpha_c = \alpha$ but $\beta_c \neq \beta$ across counties thereby using the combined data and interacting X with county dummy variables. Or, perhaps it is reasonable to consider all parameters as being the same across counties and thereby allowing the combined data to be analyzed without concern for county-specific effects.

The assumptions in this example are overly restrictive; for example, if it is reasonable to assume that regression equations have common parameters (as was done above), then, with enough data to appeal to large-sample results, one can often combine the data and use method of moments or generalized method of moments to estimate parameters without requiring the distributions be from the same family. In other words, it might be that $F_c \neq F_d$ for arbitrary pairs c and d of counties, but $E(Y|X) = \alpha + \beta \cdot X$ for all counties (or with at least one of the parameters a constant).

The main point of the preceding problem is that you started with separate data generating processes that imply different distributions for random variables, without further assumptions there is no justification for combining the data. However if it is reasonable to make certain assumptions, then common parameters may be identified that warrant using all the data in estimation. The standard errors for these estimators will be the typical ones calculated for fixed effects models.

Inherent Fixed Units, Random Effects, and Standard Errors

The preceding problems may lead you to conclude that random effects cannot be applied when a data generating process includes inherent fixed units. But this conclusion is mistaken. Random effects can be applied in such cases; however, you must be careful and understand what the probability space underlying the random effects is modeling and construct standard errors accordingly.

Problem 6.9

Consider Problem 6.8 again. You have an equal probability sample of residents (with replacement) from each of the 62 counties in New York. You are interested in the relationship between individual-level variables (i.e., measurements on the people of each county). Can you meaningfully treat the inherent fixed-unit county effects as a random effect?

Solution: We have observations from all 62 New York counties, and the counties do not compose a sample of counties. As discussed above, in terms of sampling uncertainty, the county-level should not be modeled as a random effect based on sampling probabilities; if it is, it is trivial.

However, suppose we can set up the problem, as above, in terms of the model $Y \sim F(\alpha_c + \beta \cdot X, \sigma^2)$ for some distribution F. The α_c parameters are fixed for each county, but across the counties there is a population distribution of this parameter and we can use a probability space to model this distribution. In other words, since each county has its own α_c, there must be a mean α_c across the counties and a likely variation across the α_c's, and, in general, some distribution of α_c across the population of counties. In this case, it is helpful to use $\Omega \times C$ as our outcome set in which Ω denotes the population of New York and C denotes the set of counties in New York. We can then define $(\Omega \times C, \mathcal{A}_\Omega \otimes \mathcal{A}_C, P)$ as our probability space. Here we describe P as $P = P(A_\Omega | A_C) \cdot P(A_C)$ in which $P(A_C)$ is defined to produce a model of the normalized histogram of county characteristics (such as α_c), and $P(A_\Omega | A_C)$ is modeling the patient data generating process within each county. Consequently, α_c can be considered a latent variable with a distribution associated with $P(A_C)$ as a population model of counties, and we can estimate its parameters, such as the mean and variance of the fixed-unit effects across the counties. For example, we might base our analysis on the mixture model

$$f(y|x; \beta, \theta) = \int f(y|x, \alpha; \beta) \cdot f(\alpha; \theta) \cdot d\alpha \qquad (6.110)$$

This model treats α (our county effect) as a random variable in which $f(\alpha; \theta)$ is its density associated with $P(A_C)$. After specifying our distributional families, we can estimate the model parameters, including θ, which describe the distribution of the fixed-unit effects across the counties.

We want our standard errors to only reflect sources of sampling uncertainty, and the standard errors must therefore be constructed so as not to include the variation related to the population model $P(A_C)$ as if it were sampling variation. For some models, it might be possible to work out an equation for the standard error or a Taylor series approximation of it, but it might be easier to bootstrap the standard error in which bootstrap samples are taken from within each county (i.e., a county-stratified bootstrap sampling method). In this way, we take into account the variation due to the within-county patient data generating process, but do not inappropriately account for across-county variation as if it were due to sampling. Such models can be very useful when eliminating the fixed effects is not possible and there are too many parameters to directly model. In such a case, the preceding random effects approach reduces the number of parameters from one fixed effect per county to only those describing the distribution of fixed effects.

For statistical software that allows stratified bootstrapped standard errors, estimating this model is quite easy: assuming the bootstrap algorithm is appropriate for the within-county patient data generating process, simply run a random effects model with county random effects using county-stratified bootstrapped standard errors. The results are directly inter-pretable, and the usual random effects assumptions apply.

Treating Fixed Effects as Random

In Problem 6.4, above, we declared that fixed effects are certainly possible for the classic nested data generating process. However, Problem 6.9 admits a modification, presented below in Problem 6.10, that allows us to treat such fixed effects, which are not due to inherently fixed units, as random effects that are not modeling the nested sampling design.

Problem 6.10

You collect a random sample of 1000 hospitals in the United States and then sample patients from each of those hospitals. Can you legitimately specify a model of individual characteristic y as a function of characteristic x with an intercept as a random effect modeling the distribution of the sampled hospi-tals specific effects?

Solution: Yes. You can condition on the hospitals in your sample yet model the normalized histogram of these specific hospital effects as if it were a population model. The approach is identical to that presented for Problem 6.8; however, the interpretation is different.

As mentioned above, in this approach the standard errors must only account for the within-hospital variation; however, since the distribution of hospital-specific effects is a distribution across the hospitals in the sample only, the corresponding parameters are to be interpreted accordingly. The advantage of this strategy, rather than merely using a fixed effects model, is to allow for a distribution of such effects when there are too many fixed effects to include in the model and elimination of the fixed effects is not possible, or when one is interested in the association of the outcome with hospital-level characteristics.

Conclusion

Note that the nonstandard random effects models presented in this chapter are subject to the same assumptions required for identification that apply to the standard random effects models. However, in testing these assumptions, it is important to use the correct standard errors for the nonstandard models. For example, if you run your software's usual Hausman test for random effects, your software is likely to mistakenly assume that the distribution of random effects is reflecting sampling variation. The proper standard errors for specification tests must be based on the sampling variance alone, just as the other standard errors discussed above.

Additional Readings

In this chapter, I focused primarily on problems derived from various uses of fixed effects and random effects models. I did not, however, focus on estimation of these models. Estimation procedures can be found across the many fields of applied research. For example, in econometrics, see Greene's book *Econometric Analysis*, seventh edition (Prentice Hall, 2012), Cameron and Trivedi's book *Microeconometrics: Methods and Applications* (Cambridge University Press, 2005), and Davidson and KacKinnon's book *Econometric Theory and Methods* (Oxford University Press, 2004) for just a few out of many books that address these models.

These models are also addressed in the literature on hierarchical models and multilevel models. Of the numerous books, see for examples of introductory books, Raudenbush and Bryk's *Hierarchical Linear Models: Applications and Data Analysis Methods* (Sage, 2002) and Snijders and Bosker's *Multilevel Analysis: An Introduction to Basic and Advanced Multilevel Modeling*, second edition (Sage, 2011). Also, random effect models are a type of mixed model and that literature addresses estimation as well. See, for example, Demidenko's book *Mixed Models: Theory and Applications* (Wiley, 2004).

Note, however, that each of the above referenced books as well as other books that I am aware of treat these models presupposing a cluster (nested) data generating process. Consequently, the problematic issues discussed in this chapter are not typically addressed in these books, and therefore the proper interpretation and standard errors of model parameters and estimators are not necessarily as presented in these references and related literature if the underlying probability measure is not solely representing variation in a data generating process.

*Bibliography**

Bear, H. S. (1997). *An introduction to mathematical analysis*. New York: Academic Press.

Billingsley, P. (1995). *Probability and measure* (3rd ed.). New York: Wiley.

Cameron, A. C. & Trivedi, P. K. (2005). *Microeconometrics: Methods and applications*. Cambridge, UK: Cambridge University Press.

Davidson, J. (1994). *Stochastic limit theory: An introduction for econometricians*. Oxford: Oxford University Press.

Davidson, R. & MacKinnon, J. G. (2004). *Econometric theory and methods*. New York: Oxford University Press.

Demidenko, E. (2004). *Mixed models: Theory and applications*. Hoboken, NJ: Wiley-Interscience.

Dhrymes, P. J. (1989). *Topics in advanced econometrics: Probability foundations*. New York: Springer-Verlag.

Eagle, A. (Ed.). (2011). *Philosophy of probability: Contemporary readings*. New York: Routledge.

Earman, J. (1992). *Bayes or bust?: A critical examination of Bayesian confirmation theory*. Cambridge, MA: MIT Press.

Gillies, D. (2000). *Philosophical theories of probability*. New York: Routledge.

Greene, W. H. (2012). *Econometric analysis* (7th ed.). Boston, MA: Prentice Hall.

Hall, A. R. (2005). *Generalized method of moments*. Oxford: Oxford University Press.

Kolmogorov, A. N. & Fomin, S. V. (1970). *Introductory real analysis*. New York: Dover.

Kyburg, H. E., Jr. & Thalos, M. (Eds.). (2003). *Probability is the very guide of life: The philosophical uses of chance*. Chicago, IL: Open Court.

Mayo, D. G. (1996). *Error and the growth of experimental knowledge*. Chicago, IL: University of Chicago Press.

Mayo, D. G. & Spanos, A. (2011). *Error and inference: Recent exchanges on experimental reasoning, reliability, and the objectivity and rationality of science* (1st paperback ed.). New York: Cambridge University Press.

Mellor, D. H. (2005). *Probability: A philosophical introduction*. New York: Routledge.

Raudenbush, S. W. & Bryk, A. S. (2002). *Hierarchical linear models: Applications and data analysis methods* (2nd ed.). Thousand Oaks, CA: Sage.

Resnick, S. I. (1999). *A probability path*. Boston, MA: Birkhauser.

Schervish, M. J. (1995). *Theory of statistics*. New York: Springer-Verlag.

Snijders, T. A. B. & Bosker, R. J. (2012). *Multilevel analysis: An introduction to basic and advanced multilevel modeling* (2nd ed.). Los Angeles, CA: Sage.

Spanos, A. (1999). *Probability theory and statistical inference: Econometric modeling with observational data*. Cambridge, UK: Cambridge University Press.

Sprecher, D. A. (1970). *Elements of real analysis*. New York: Dover.

---

* This bibliography provides the full citations for the works presented in the Additional Readings sections of each chapter.

Stoll, R. R. (1979). *Set theory and logic*. New York: Dover.

Suppes, P. (2002). *Representation and invariance of scientific structures*. Stanford, CA: CSLI.

Taper, M. L. & Lele, S. (2004). *The nature of scientific evidence: Statistical, philosophical, and empirical considerations*. Chicago, IL: University of Chicago Press.

Index

Printed and bound by CPI Group (UK) Ltd, Croydon, CR0 4YY

24/10/2024

01778279-0019